Crosslinking and Scission in Polymers

NATO ASI Series

Advanced Science Institutes Series

A Series presenting the results of activities sponsored by the NATO Science Committee, which aims at the dissemination of advanced scientific and technological knowledge, with a view to strengthening links between scientific communities.

The Series is published by an international board of publishers in conjunction with the NATO Scientific Affairs Division

A Life Sciences **B Physics**	Plenum Publishing Corporation London and New York
C Mathematical and Physical Sciences **D Behavioural and Social Sciences** **E Applied Sciences**	Kluwer Academic Publishers Dordrecht, Boston and London
F Computer and Systems Sciences **G Ecological Sciences** **H Cell Biology**	Springer-Verlag Berlin, Heidelberg, New York, London, Paris and Tokyo

Series C: Mathematical and Physical Sciences - Vol. 292

Crosslinking and Scission in Polymers

edited by

O. Güven

Department of Chemistry,
Hacettepe University,
Ankara, Turkey

Kluwer Academic Publishers

Dordrecht / Boston / London

Published in cooperation with NATO Scientific Affairs Division

Proceedings of the NATO Advanced Study Institute on
Advanced Methods of Determination of Crosslinking and Scission in Polymers and their Effects on Mechanical Properties
Kemer-Antalya, Turkey
September 4–16, 1988

Library of Congress Cataloging in Publication Data

NATO Advanced Study Institute on "Advanced Methods of Determination of Crosslinking and Scission in Polymers and their Effects on Mechanical Properties" (1988 : Kemer, Kemer Bucağı, Antalya İli, Turkey).
Crosslinking and scission in polymers : proceedings of the NATO Advanced Study Institute on "Advanced Methods of Determination of Crosslinking and Scission in Polymers and their Effects on Mechanical Properties" / edited by O. Güven.
p. cm. -- (NATO ASI series. Series C, Mathematical and physical sciences ; no. 292)

1. Polymers--Mechanical properties--Congresses. 2. Crosslinking (Polymerization) 3. Scission (Chemistry) I. Güven, O (Olgun) II. Title. III. Series.
TA455.P58N378 1988
620.1'9204292--dc20 89-28176

ISBN-13: 978-94-010-7359-2 e-ISBN-13: 978-94-009-1924-2
DOI: 10.1007/978-94-009-1924-2

Published by Kluwer Academic Publishers,
P.O. Box 17, 3300 AA Dordrecht, The Netherlands.

Kluwer Academic Publishers incorporates the publishing programmes of D. Reidel, Martinus Nijhoff, Dr W. Junk and MTP Press.

Sold and distributed in the U.S.A. and Canada by Kluwer Academic Publishers,
101 Philip Drive, Norwell, MA 02061, U.S.A.

In all other countries, sold and distributed by Kluwer Academic Publishers Group,
P.O. Box 322, 3300 AH Dordrecht, The Netherlands.

Printed on acid free paper

CONTENTS

MAIN-CHAIN SCISSION OF LINEAR POLYMERS AS STUDIED BY PULSE RADIOLYSIS AND FLASH PHOTOLYSIS

W. Schnabel

CARBON 13 NMR OF GAMMA-IRRADIATED POLYETHYLENES

J. C. Randall

THE USE OF PULSED NMR TO DETERMINE NETWORK FORMATION BY CROSSLINKING AND ENTANGLEMENT

A.Charlesby

METHODS FOR STUDYING THE EFFECTS OF HIGH-ENERGY IRRADIATION ON THE MECHANICAL AND ELECTRIC PROPERTIES OF POLYMERS
P. Hedvig

THERMALLY STIMULATED DISCHARGE AND DIELECTRIC MEASUREMENTS OF CURE IN RESINS AND GELLED MATERIALS
R. A. Pethrick

ULTRASONIC AND VISCOELASTIC STUDIES OF CHAIN ENTANGLEMENT IN POLYMER SOLUTIONS
R. A. Pethrick

PREFACE

Crosslinking of polymers has tremendously extended their range of applicability in the past two decades and today end-use properties of polymers depend critically on the extent and distribution of crosslinking. As of mid-seventies there were no direct chemical methods available for the determination of crosslinks in polymers. Measurements were based on the consequences of crosslinking reactions and yielded lower limits of crosslinking. There have been important new developments in the techniques of direct and absolute determinations of crosslinking not only measuring the number of crosslinks but also extent of physical entanglements.

It was the objective of this NATO Advanced Study Institute on " Advanced Methods of Determination of Crosslinking and Scission in Polymers and their Effect on Mechanical Properties" to provide a critical analysis of the power and limitations of various techniques. This volume contains the main lectures given at the NATO ASI held in Kemer-Antalya, Turkey, September 4 - 16, 1988. I hope that with the expert contributions of the lecturers of this ASI, the researchers will be provided with an up-to-date survey of presently available techniques and alternative ways of using conventional methods of analysis in the determination of crosslinking and scission in polymers.

This NATO Advanced Study Institute was made possible by a grant from the NATO ASI programme of the Scientific Affairs Division of NATO. In the name of lecturers and the participants of this ASI, the editor of this volume gratefully acknowledges the sponsorship of NATO. I also would like express my appreciation to the U. S. Army Research, Development and Standardization Group (U. K.) for their interests and support on this meeting. Finally I would like to thank to my students M. Şen , D. Şolpan, A.Güner and F. Yiğit for their helps in the preparation and rewriting of some of the chapters.

Olgun Güven
July 1989, Ankara

MOLECULAR WEIGHT CHANGES AND NETWORK FORMATION BY SCISSION AND CROSSLINKING

A . Charlesby
Silver Spring, Watchfield,
SWINDON SN 6 8 TF, UK

INTRODUCTION

Many of the physical properties of a polymer can be profoundly modified by changes in molecular weight due to main chain scission, resulting in a reduction in molecular weight, or by crosslinking which at first results in a slight increase in molecular weight, and then by the formation of a network, which can result in a complete change in mechanical behaviour.
Network formation can be produced by chemical means, or by radiation which has great advantages for quantitative analysis and study of resultant physical behaviour.
Outlines are given of several possible methods of providing such network structures, and the nature and extent of change. These changes can result not only in a chemical modification, but also in a profound alteration in polymer morphology.

MAIN CHAIN SCISSION OF POLYMERS

Main chain scission of long chain polymers (degradation, which does not imply any social or moral lapses) is to be distinguished from depolymerization although both involve a reduction in average molecular weight, and corresponding change in physical properties. The conditions understood by main chain scission involve a random distribution along molecular length, so that whatever the initial molecular weight distribution, it rapidly tends towards a random molecular weight distribution, with the weight average M_w and "z" average M_z being twice and three times the number average M_n. Where scission would not occur at random, a very different set of conditions may prevail, but there is little published evidence of this except in a multiphase polymer (partially crystalline) or a copolymer.

O. Güven (ed.), Crosslinking and Scission in Polymers, 1–13.

A number of important polymers including polyisobutylene (PiB), Teflon (PTFE), cellulose and poly methyl methacrylate (PMMA) show this behaviour. If $M_{n,o}$ is the initial number average M_n ,it will decrease to a value $M_{n,r}$ after a dose of r kGy of high energy radiation, where

$$1 / M_{n,r} = 1 / M_{n,o} + 1.04 \times 10^{-4} G(s) r$$

where G(s) represents the number of scission per 100 ev absorbed. This is often about 1 - 2, and varies little if at all with molecular weight over a very wide range. It can therefore be related to the intrinsic viscosity [η]

$$[\eta] = K M_n^a = K (M_w/2)^a$$

for a random distribution Plotting log [η] or any other measure of molecular weight against log r should therefore give a linear plot from which G(s) can be deduced.
This simple expression leaves open a wide range of questions which will be presented briefly.

(1) Fracture of the main chain leaves two radicals at the side of scission. Which chemical changes then occur to stabilise these highly reactive ends ? They can be detected by E.S.R. at low temperatures, and their rate of disappearance at various temperatures may provide evidence of molecular mobility in the transition form crystal or glass to liquid on a molecular scale.

(2) Side chains can also be broken off. Does this occur at the side of main chain scission, or do they occur as independent reactions anywhere along the chain? Here again we need a closer ESR study. The short molecular fragments may react further within what may be a solid matrix, or migrate away.

(3) Is there some form of chain repair, so that not all initial scissions become permanent? Evidence for some form of repair is shown by polyisobutylene, where G(s) rises from 2.2 at -196 °C to 10 at 90 °C, see e.g. the paper by Alexander and Charlesby. Presumably at the higher temperature the initial chain ends can move apart more readily, or react and stabilize . For PMMA the corresponding values are 0.5 to 2.2, again an increase of 4.5 times.
It is highly significant that the same behaviour is shown in a number of bacteria when irradiated in dry form . This would indicate that radiation - induced inactivation in such systems is associated with main-chain breaks or some closely related radiation-chemical reaction, and is not at all specific to biological systems. Indeed one may enquire as to the parallel behaviour of

biopolymers and polymers under irradiation.

(4) In wiew of the similarity in scission behaviour as between PiB (a rubbery material at room temperature) and PMMA (a glass) how do the broken ends migrate away from each other in the glassy state. This poses an interesting question on molecular mobility in the frozen state, which would of great interest to follow by the changes in radical nature and concentration with time after irradiation, using E.S.R techniques for example .

It is interesting to note that in glassy PMMA gas bubbles form if an irradiated specimen is subsequently warmed, allowing gaseous fragments to move together. Are these fragments held separately or do they form from the polymer radicals only when the latter acquires adequate mobility ?

Another interesting feature is that these bubbles do not form near the surface of a piece of PMMA. It must be surmised that this distance of about 1-2 mm is the distance for agglomeration and gases formed near the surface can escape. Is there some form of selection or segregation according to gaseous molecular size or nature? A study of this system under a variety of conditions e.g. different dose, giving different occluded bubble concentration, or different temperatures below but near the glass temperature.

(5) PTFE is a largely crystalline polymer at room temperature, and irradiation causes main chain scission and a small increase in degree of crystallinity. One may then enquire as to whether scission occurs equally readily in the crystalline and in the amorphous regions which separate them . One would expect a considerable difference though this is not seen in the G (scission) values for PiB and PMMA.

(6) If irradiated PTFE is subsequently heated above its melting point, and then recrystallised, there is a considerable increase in the density and degree of crystallisation. It can be assumed that many molecular chains can take part in several crystalline regions and therefore keep them apart. When scissioned in the amorphous regions, these shorter chains no longer participate in several crystals , which can therefore grow further without this form of interference. Main chain scission can therefore serve as a method of investigating crystallization reactions.

(7) An extremely important aspect of main chain scission is in radiobiological work involving DNA. Single strand breaks occur naturally and can be repaired, but in the helix form, to become permanent both strands must be broken, and presumably in close proximity. With the radiation doses involved this is extremely unlikely and another explanation must be sought allowing some form of correlation between such scissions. One suggestion is that energy is imparted from high energy radiation along short secondary tracks. If each causes several ionizations/excitations along its very short

trajectory some scissions may be found in close proximity. However as yet there is no evidence of such track effects in simple polymers, although the interaction with incident radiation would be very similar. The location of adjacent but separate ionisation-excitation events, due to the passage of a single high-energy particle is of considerable interest in radiobiology.
The nature and characterization of these single and double strand breaks in DNA is of fundamental importance in radiobiology and cancer research and considerable information might be transferred from the simple but related work on scission in long chain polymers .

CROSSLINKING

The physical behaviour and especially the mechanical properties of long chain polymers depend not only on chemical structure and chain mobility (and therefore on temperature), but also on the association between adjacent molecules. This can take the form of crystallinity, rigidity as in the glassy state, chain entanglements, hydrogen bonding etc. An extremely important form of such intermolecular binding is by crosslinks, representing permanent chemical bonds between what were originally separate molecules. They can occur either directly from one polymer chain to another such as by C-C bonds, or indirectly via a short chemical chain as in the S bonds in rubber vulcanization.
The effect of such bonds depends not only on their density and on the chemical structure of the individual bonds, or even on the molecular mobility and morphology, but also on the distribution of these crosslinks . They may be distributed at random, they may be random except in special regions e. g. crystalline regions, or they may be correlated- i. e. if polymer A is linked to polymer B near a point P, there is an above-average chance of B being linked to polymer molecule C near the same point P. The distribution of crosslinks must also be compared in the crystalline regions with those in the more flexible, amorphous regions. Suitable additive (fillers) such as carbon in rubber may also play an important role. The theory of high elasticity is well-known and is dependent on a statistical approach to molecular arrangement, with corrections for mutually excluded volume, chain rigidity, end effects etc. The theory of tensile strength is less well-understood, and the transition from highly elastic to glassy behaviour can be traced quantitatively for the identical polymer, merely by extending the radiation dose over perhaps a 50 fold range.

Many of these physical properties dependent on crosslink density are also influenced by chain entanglements , which can behave for limited periods as equivalent to permanent crosslinks. The mechanical behaviour of a crosslinked and entangled network will be a composite of permanent elongation plus flow, the latter being due largely to the changes in entanglements under stress. This opens an intriguing possibility of analysing the mechanical behaviour of such crosslinked and entangled networks in terms of their density of links.

Crosslinking can be produced in long chain polymers by a number of chemical reagents, but also by high energy radiation, which has a number of advantages; the reaction shows relatively little dependence on temperature , it leaves no chemical residues similar to those left by chemical catalysts , it can be induced over an enormous range of intensities and ranges, so that the reaction can be studied in the greatest detail at any stage . Furthermore many of these crosslinking reactions are in direct proportion to the dose , so that a quantitative analysis can be achieved, and this allows a direct link between crosslink density and physical properties .

RANDOM CROSSLINKING

The simplest form of crosslinking is one where two radicals on adjacent molecules are linked together to form a direct crosslink . The first change is merely an increase in molecular weight, as when two linear molecules are linked at random to form an X type of structure (with unequal lengths of course). This is shown by a small increase in viscosity, although the weight average M_w , if measured by light scattering, will show a greater change.
A drastic change occurs when there is an average (δ) of one crosslinked unit (=0.5 crosslink since each crosslink requires a crosslinked unit on each of the chains) per weight average molecule M_w . At this concentration, termed the gel point, there is an incipient three dimentional network , whose properties are inherently different form the original polymer . This network is essentially a single three-dimensional molecule, and one cannot therefore say that it is a solid, liquid or gas. With increasing radiation dose or degree of crosslinking, this network (or gel) forms an increasing (gel) fraction of the total, while the residue of soluble molecules (sol) shrinks rapidly. A mathematically important relationship exists between this residual soluble fraction s and the crosslink density. This may be expressed in terms of the average number of crosslinks δ per weight average molecule , or when this average becomes very large and the sol fraction or ends of each chain is almost negligible, it becomes more convenient to express it in terms of the average molecular weight between successive links M_c .
Several methods are available to determine the crosslink density under these simple conditions of a random distribution, with no interfering effect such as partial crystallinity and direct proportionally to dose.

(1) Determination of gel dose r_g (by extrapolation of the soluble fraction s,

down to s =1. Then δ=1 whatever the molecular weight distribution , for a random crosslink distribution

(2) Determination of sol fraction in relation to dose r. This will depend on the molecular weight distribution i.e. weight average M_w , z average , M_z, ... The general expression is

$$g = \delta g - \frac{1}{2!} \delta^2 g^2 \frac{M_z}{M_w} + \frac{1}{3!} \delta^3 g^3 \frac{M_4 M_z}{M_w^{\,2}} - \cdots$$

where δ is the average number of crosslinked unit per weight average M_w (g = 1 - s , where s is the soluble fraction) .
If the initial distribution is uniform

$$s = \exp (- \delta \sqrt{1 - s})$$

while if it is random

$$s + \sqrt{s} = 2 / \delta$$

and if it is pseudo random ($M_w = M_z/2 = M_4/3 = \cdots M_i / (i - 1)$)

$$s = 1/ \delta$$

where even a very slight degree of crosslinking initiates formation of a network.

(3) Swelling of the network
For a completely crosslinked system

$$\ln (1 - x) + x + \mu x^2 + \rho \upsilon (x^{1/3} - x/ 2) \rho\upsilon / M_c = 0$$

Where ρ is the polymer density, υ is the molar volume of the solvent, and x is the original volume / swollen volume, or for large degrees of swelling (with V = 1/x = swelling ratio)

$$V^{5/3} = (0.5 - \mu) M_c / \rho \upsilon$$

so that once the constants are known, M_c can be deduced directly from the swelling ratio V. In this relation any soluble fraction or chain end effects are ignored.

(4) If the specimen is well above the glass temperature or crystal melting point, the crosslinked network acquires highly elastic properties, and the elastic modulus E is inversely proportional to the crosslink density or its reciprocal M_c .

$$E = 3\rho RT / M_c$$

with ρ=density , RT are gas constant and absolute temperature , a correction should be applied for chain ends (i.e. the portion of the chains containing only one linked unit, and therefore not constituting part of the elastic network). The value of this correction depends not only on M_c but also on the molecular weight distribution, and may be very considerable until $\delta >> 2$.

(5) More recent evidence has been found showing that pulsed NMR can be used to distinguish between network (gel=1 - s) and soluble fraction s, and also provide measures of M_c. This will be discussed separately.
By following the changes in mechanical properties with crosslink density over a very wide range-as can be achieved by radiation one can pass through the following physical changes

(a) At low crosslink densities ($\delta < 1$) the polymer is completely soluble. At low temperature it may be somewhat flexible if it is partly crystalline, or rigid as in a glassy state. At high temperature it becomes a viscous fluid, though it may retain a shape for short periods due to chain entanglements.

(b) At higher crosslink densities ($\delta > 1$) it is almost unchanged in mechanical properties while still crystalline or glassy but at higher temperatures it takes on rubber-like properties of high elasticity. The modulus is proportional to radiation dose i.e. crosslink density. The tensile strength also increases.

(c) At even higher crosslink densities, the elastic modulus continues to increase, but the properties are successively those of an overcured rubber , a cheesy, leathery and finally glassy state. The transition between these states can be directly related to the radiation dose so that the radiation technique offers an excellent quantitative method of relating elastic properties to M_c. The theory of this dependence is only known for relatively low densities of crosslinking, when the polymer behaves as a crosslinked rubber. The modulus at much high crosslink densities even with the glassy state, deserves fuller attention.The polymer chains in a rigid (crystalline or glassy) state are bound in position, and can show only local changes in arrangement in reasonable times. Above the glass or crystallisation temperature, the chains may be held together largely by permanent crosslinks and temporary entaglements. For short deformation times both behave as crosslinks, and the flexible polymer can behave as a fully crosslinked system. For longer periods under stress, some molecular motion allows the entanglements to disappear, to be reformed elsewhere, giving rise to behaviour formed flow . There will be an elastic memory due to the permanent radiation-induced crosslinks, and by a series of combined measurements of flow and recovery for various times and radiation doses, the rheological behavior of such polymers should be studied.

ENHANCED CROSSLINKING

In many crosslinking reactions, there is a direct proportionality between the number of crosslinks and the chemical reagents or better the radiation doses. This is best illustrated by the radiation dose r_g (expressed in kGy) needed to form an incipient network with a random distribution

$$G\, r_g\, M_w = 0.47 \times 10^7$$

where G (the number of crosslinked units per 100 ev absorbed) and M_w is the weight average molecular weight. Typically $G \approx 1$, so that if $M_w = 5 \times 10^5$, $r_g = 9.4$ kGy approximately. In the case of polystyrene G is very much smaller for radiation chemical reasons, but much higher values are very unusual.

For a series of polyfunctional monomers, or unsaturated polymers, a high degree of crosslinking, giving a rigid or glassy structure , is obtained at far smaller doses, even if M_w is low , say 10^4 . Obviously a very different crosslinking system is involved, with a G value for crosslink formation amounting to perhaps one-hundred times that expected from the energy input. This behaviour is termed enhanced crosslinking, and is in fact a combination of crosslink and a polymerization or grafting reaction. It requires multiple unsaturation either in the polymer P itself, or in a low molecular weight additive known as a sensitizer, S

What is involved is a reaction of the following character

P —> P· + etc. or S —> S· + ?

P· + S —> PS·

PS · + P —> PSP·

PSP · + S + P —> PSPSP · etc . or variants

A single initation step can therefore result in a number of crosslinks associated with each other in close proximity. The relation between network formation and radiation dose is quite different from that applicable to random crosslinking, as outlined above. In particular the network first

begins to appear at a greatly reduced radiation dose or initiating chemical concentration. This is the behaviour found in unsaturated polyesters for example, and is also present to some extent in polyisobutene.
Since this is in part a polymerization reaction proceeding through a series of polymers, some means must be found to terminate the reaction, once commenced. This can be achieved for example by the combination of two such crosslinking chains. If i is the chance of a given link being the last one (i.e. reciprocal of the number of links per initiating species) and $i << 1$; $\delta_g \approx i/2$ as against 1 for random crosslinking. The required radiation dose r_g, is now

$$G\, r_g\, M_w \approx i^2\ 0.47 \times 10^7$$

Then the value of G(crosslinks) at the gelation dose appears to have increased by a factor of the order of i^2. The plot of sol/dose curve is also very different.
The use of irradiated linear polymers such as polyethylene and rubber is usually to enhance their radiation stability at higher temperatures, or to make use of the memory effect. For the polymers utilised in enhanced crosslinking it is usually to obtain a much higher degree of crosslinking with a very modest dose, though sometimes at the cost of a more expensive sensitizing monomer.
The relationship between such highly crosslinked systems and the resultant physical properties has not been studied to the same theoretical extent as have those for lightly crosslinked polymers showing high elasticity. The transition from a very lightly crosslinked polymer, through maximum tensile strength until eventually a glass-like state is reached, deserves far greater attention.
One may enquire why poly - trans isoprene shows this tendency towards enhanced crosslinking, though it is absent in the equally unsaturated polyisoprene. One possibility is that for a chain crosslinking reaction to occur there must be a suitable morphological arrangement of adjacent molecular chains and this is possibly present in the trans configuration , whereas with the cis form links between polymer molecules A and B place A and B locally in a geometric arrangement where they cannot approach adjacent molecules to allow crosslinking.

The use of so-called sensitizers, notably trifunctional unsaturated monomers, to crosslink linear polymers which would otherwise be resistant to this reaction-e.g. P.V.C., may involve a link from one polymer molecule to another via the sensitizer, but the doses involved appear too low. Another possibility is that the sensitizer forms a three-dimensional network very readily, and that the polymer is trapped within its pores, appearing insoluble as a result. Measurements on low molecular weight PVC, polystyrene etc. might answer this important question. Another matter for study is how these polyunsaturated systems build up a three-dimensional network so readily. A simple polymerization reaction should not, at the low doses involved, do more than form largely linear polymerization chains.

The use of radiation to promote such enhanced crosslinking is becoming of increasing interest, although the basic mechanisms are not clearly understood. It differs from simple polymerisation which results in a linear reaction, with little or no build-up of network except at higher doses. To produce the same degree of crosslinking without the polyunsaturated monomer would necessitate doses of several hundred Megarads. Furthermore the reaction proceeds with a low molecular weight monomer, and to an extent which depends on dose, but not on dose rate. This would imply that each initation (radical or ion) gives a three-dimensional sequence, forming a dense network in its neighbourhood, or that this sequence terminates within the vicinity of the initiator, possibly due to exhaustion of reactive monomer. Only a limited amount of such individual zones overlap to give overall properties .
The rapid curing of such polyfunctional monomer plus polymer, using radiation, has a number of advantages as compared with the chemical method, including purity (no residues from initiator) very rapid reactions even at low concentration, ready control of the degree of reaction etc. This technique is acquiring increasing intrest in industry, especially for thin films and low energy radiation, electrons or even ultraviolet, the ability to control the degree of reaction with depth, or even with shapes, could offer very interesting new possibilities.

OTHER FORMS OF CROSSLINKING

A number of possibilities occur in the formation of a network. One of these is by H bonding, which involves a special chemical structure. Another is termed endlinking, where a scissioned molecule can, via a reactive radical end, link to a linear molecule to give a T type of intermolecular bond. It is assumed that of all polymer molecules scissioned , a fraction α can form endlinks. In this case the sol fraction depends on α

$$s = (1 - \alpha)^2 / 9\alpha^2$$

so that for gel formation ($s < 1$), α must exceed 0.25 . The relation for network formation with $\alpha = 0.25$

$$G\, r_g M_w = 0.64 \times 10^7 \quad (\text{ r in kGy })$$

as compared with

$G\, r_g\, M_w = 0.47 \times 10^7$

for simple random crosslinking.
The full plot of soluble fraction s versus dose for endlinking is not very different from that for random crosslinking, and the two may well be confused.

COMBINED CROSSLINKING AND SCISSION

In spite of the random absorption of energy by irradiated polymers, there is a very sharp distinction between those which crosslink and those which scission. It must therefore be accepted that a number of steps intervene before the final chemical changes, steps which involve selective bond rupture and formation (usually) of reactive radicals.
To see how far the two processes take place on a polymer, one can most conveniently use the Charlesby - Pinner relation $s + \sqrt{s} = 2 / \delta = 2\, r_g / r$, which is strictly only applicable to polymers, initially of a random molecular weight distribution ($M_w = 2\, M_n$), and for a random distribution of crosslinks. Then a plot of $s + \sqrt{s}$ versus r^{-1} should give a straight line passing through the origin (s = 0) and r_g. This simple relation has been widely used, often where it is inappropriate, instead of the more general relations for other initial distributions. However if crosslinking and scission occur in the same polymer, and both in proportion to dose (with crosslinking predominating), the distribution will rapidly become a random one and an approximately straight line plot of $s + \sqrt{s}$ is still obtained, but extrapolating not to 0 (when r_g —-> infinity) but to a higher value of s or $s + \sqrt{s}$ This extrapolated value of s is a direct measure of the ratio of main chain scission to crosslinks. This is notably the situation for polypropylene intermediate between the crosslinking polyethylene and the scissioning polyisobutylene.

ANTIOXIDANTS

Crosslinking reactions especially in linear or branched polyolefins, can be greatly affected by the presence of oxygen, which reacts with the radicals which would otherwise provide crosslinks. This is reduced by the addition of antioxidants such as sterically hindered phenols or aromatic amines.

The effect of these and other antioxidants may not only reduce the oxygen effect,it may also reduce the degree of crosslinking, so that higher doses are needed for the same degree of crosslinking.This could be by the transfer of energy to the antioxidant, or the scavenging of polymer radicals which might otherwise result in a crosslinked species. Another question to determine is whether in either of these two patterns of behaviour, the antioxidant molecule is itself modified and can no longer take up this function .

As an example one may quote a LDPE polyethylene whose crosslinking is enhanced by about 35 % using a sensitizer. This sensitizing effect is reduced to 16 % by only 0.25 % of an antioxidant to 10 % with 0.5 % and to 6 % with 1%.

FILLERS

The introduction of very small particles such as carbon into highly elastic networks can profoundly influence its mechanical properties, and there are several theories available to explain this behaviour. One assumes that additional crosslinks are formed between carbon particles and rubber for example, but such additional links could also result from an increased dose in the absence of such particles, and this does not appear to be the case. Other formes of intervention can also be considered, but the one favoured by the author is a largely physical one, in which the particles are embedded into the pores of the rubber network and therefore reduce its deformability rather like a fishing net has its flexibility and deformability reduced if there are fishes trapped with the pores.

CROSSLINKING OF POLYMERS IN SOLUTION

The irradiation of biopolymers in an aqueous environment has received considerable attention because of its interest in radiobiology. The parallel, but simpler model of a polymer in water shows many features which may help to interpret the radiobiology work, but which are of great interest in their own right.

Many water-soluble polymers can be crosslinked by radiation, even though the individual molecules may be some distance apart. The irradiated polymeric system can become crosslinked above a gelation dose r_g, forming a network swollen with water. At first the amount of swollen polymer increases with dose above r_g, and then decreases approximately as $r^{0.6}$ as required by the conventional swelling theory. The earlier rise is of course due to the swelling of only a fraction of the polymer, namely that fraction (gel) which is forming the network .

The second feature is that the lower the polymer concentration, the lower is the dose needed to form a crosslinked network. In other words the further apart the individual molecules, the easier it is to link them together. The explanation is that one is also irradiating the water molecules, and the fragments into which this broken can attack the polymer to give radicals capable of linking them together . Obviously the fewer polymer molecules, the more each one can be affected by these fragments-termed the indirect effect. The net effect is therefore due to crosslinking by direct and indirect effects.

A third feature is that below a certain concentration (typically about 1%) , it is no longer possible to form a network , even at very high doses. This is not because there is any change in the radiation-induced chemical changes. The explanation is that crosslinks can be formed by radiation, either between the separate, but neighbouring, polymer molecules (external links) or between two parts of the same molecule which happen to swim into each others proximity. Such internal links, have the same chemical nature as the external links, but simply serve to bind together parts of the same molecule to form a micronetwork. As such internal links increase with dose, this micronetwork shrinks, and therefore moves further away from other molecules.

This aspect of crosslinking of polymers in aqueous solution has not received adequate study. The formation of a water-swollen network has been utilised for medical purposes, the pores can be readily and reproducibly controlled by the radiation dose, and may act as a selective membrane .

At the lower doses, the polymer forms submicroscopic and separate structures , again with physical properties readily controlled by the radiation dose. A further feature is that these reactions can be greatly modified by the presence of very small concentrations of additives , which can react with the radiation products of water or polymer , or even with biopolymer , and the nature of these reactions can be of great interest in radiobiological work.

REFERENCES

The basic equations relating crosslink or scission density to radiation dose or radical concentration have been published in the early days of the subject of radiation of polymers, and are summarized in " Atomic Radiation and Polymers " by A. Charlesby (Pergamon 1960) which also gives the original references e. g.

Proc. Roy . Soc . (London) A (1954) 222 , 60 , 542 ; A(1954) 224 ,120 ;A (1955) 231 , 521 ; A (1957) 241 , 495.

Flory , P.J . Principles of Polymer Chemistry (Cornell) 1953 .

Alexander, P. , Black, R.M. , Charlesby, A. Proc Roy Soc A 232 31 1955

Alexander, P. , Charlesby, A. , Ross, M . Proc Roy Soc A 223 392 1954

HIGH ENERGY RADIATION- AND UV LIGHT-INDUCED CROSSLINKING AND CHAIN SCISSION

WOLFRAM SCHNABEL

HAHN-MEITNER-INSTITUT, BEREICH S
GLIENICKER STR.100
D-1000 BERLIN 39
Federal Republic of Germany

ABSTRACT. The 1st part of this chapter deals with fundamental aspects concerning (a) the absorption of UV and high energy radiation by matter, (b) the detection of crosslinks and scissions in linear polymers and (c) mechanistic aspects regarding crosslinking and main-chain scission. The 2nd part concerns some recent results regarding ion beam- and soft X-ray-induced radiation effects in linear polymers

1. Introduction

1.1 IMPORTANCE OF RADIATION-INDUCED CROSSLINKING AND MAIN-CHAIN SCISSION IN LINEAR POLYMERS

There are various modes of inducing crosslinking (XL) and chain scission (CS) in polymers, the most important ones concern thermal, mechanical, biological, chemical, radiation chemical and photochemical processes.

Characteristic for the radiation-induced modes is the following:

XL and/or CS can be initiated at low temperatures by simply placing polymer specimen into the radiation field. Moreover, XL and/or CS occur spontaneously without the aid of added low molar mass additives.

Notably, radiation is a convenient tool to alter physical properties of polymers, since XL and/or CS cause alterations in the average molar mass and many practically important physical properties depend on the molar mass. In this connection it should be pointed out that XL commonly exerts beneficial effects on mechanical properties up to a certain absorbed radiation dose, i.e. up to a certain degree of crosslinking, whereas CS leads to severe property deteriorations right after the irradiation has started. Irradiation of polymers to high absorbed doses is always harmful: it causes embrittlement and color changes. This applies both to treatment with high energy radiation and UV light. Since the sunlight penetrating the earth's atmosphere

O. Güven (ed.), Crosslinking and Scission in Polymers, 15–36.

contains a portion of UV light, photodegradation of polymers caused by XL and/or CS is of enormous practical and economical importance with respect to commercial polymers. Notably, polymer degradation is frequently significantly enhanced by oxidative processes which are initiated by radiation and perform in many cases kinetically as "chain reactions", i.e. as auto-oxidative processes involving CS. To overcome this problem stabilizers are added quite frequently to commercial polymers. Stabilizers are low molar mass compounds which protect the polymer with respect to XL and CS against the action of UV light.
Apart from the detrimental effects of radiation-induced XL and CS, there are various quite useful applications. Typical examples for the beneficial effects of Xl and CS in polymers apply to photolithographic techniques which are used to manufacture integrated circuits, compact discs, cathode ray tubes and printed circuit boards. Moreover, high energy induced XL is applied in the production of shrinkable polyethylene tubes and foils, and to improve thermal and mechanical properties of insulations of high voltage cables.
The main intention of this paper is to give a brief introduction into the field of radiation and photochemistry of polymers as far as XL and CS are concerned. For this purpose types of radiation of interest, radiation sources and the absorption of radiation by matter will be treated in the next sections. Furtheron, the detection of XL and CS and the determination of XL and CS yields will be dealt with in a concise form. At the end of the chapter some recent results obtained in the author's laboratory will be presented

1.2. TYPES OF RADIATION AND RADIATION SOURCES.

The types of radiation concerned in this article are listed in Tables 1 and 2. Electromagnetic radiations are to be discriminated from particle radiations. Commercially available lamps such as deuterium,xenon or mercury lamps serve as sources for visible and UV light. In addition, powerful lasers emitting monochromatic and coherent light are used nowadays in research work and practical applications. Quite popular are solid state lasers such as ruby and Nd-YAG systems and gas lasers producing light of wavelengths covering the range from UV to IR. X-rays can be produced by commercially available tubes. Very powerful X-ray sources are electron synchrotrons which are now operated in various countries.
Apart from the classical particle radiations, α- and β-rays, electrons as well as light and heavy ions of energies ranging up to several hundred eV can be produced nowadays by particle accelerators, and neutrons are generated in nuclear reactors.

1.3. ABSORPTION OF RADIATION

The absorption of electromagnetic radiation of wavelengths corresponding to infrared (IR), visible and ultraviolet (UV) light is

Table 1. Types of Electromagnetic Radiation

Type of Radiation	λ (nm)	E_{Photon} (eV/photon)	(kJ/mol)
	800	1.55	149.6
Visible Light	{		
	400	3.10	299.2
Ultraviolet Light	{		
	100	12.4	1194
	10	124	1.19×10^4
X-Rays	{		
	0.01	1.24×10^5	1.19×10^7
γ-Rays	≤ 0.001	$\geq 1.24 \times 10^6$	$\geq 1.19 \times 10^8$

Table 2. Types of Particle Radiation

Type of Radiation	Source
Neutrons	Nuclear Reactors
β-particles (electrons)	Radicactive Decay (Accelerators)
α-particles ($^4_2He^{2+}$)	Radioactive Decay
H^+, D^+, $^4_2He^{2+}$ and	
heavy ions:	Accelerators (van de Graaff, Linear
e.g. $^{16}_8O^{4+}$,$^{20}_{10}Ne^{7+}$, $^{40}_{18}Ar^{8+}$ etc.	Accelerators, Cyclotrons etc.)

determined by the availability of appropriate energetic levels E_2 and E_1 ($E_2 > E_1$ and $E_2 - E_1 = h\nu$) in a molecule or in a "chromophoric" group of a molecule. The absorption of light of wavelength λ by chromophoric groups at the concentration c (in mol/l) in a system of optical path length d (in cm) is described by the Lambert-Beer law

$$I = I_o 10^{-\epsilon_\lambda cd} \tag{1}$$

I_o: intensity of incident light, ϵ_λ: decadic molar extinction coefficient (in l/mol cm).

Typical absorption spectra of polymers are shown in Fig.1.

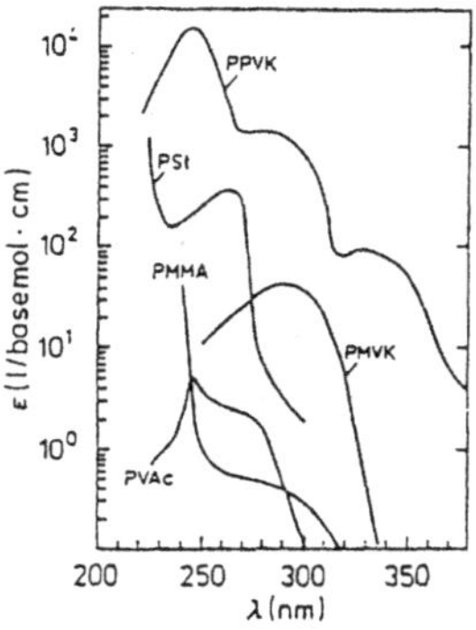

Fig.1. Optical absorption spectra of several polymers recorded in dioxane solution.
PMMA: poly(methyl methacrylate)
PVAc: polyvinylacetate
PSt: polystyrene
PMVK: poly(methyl vinyl ketone)
PPVK: poly(phenyl vinyl ketone)

Only a small number of practically important polymers is capable of absorbing solar radiation. However, quite frequently commercial polymers contain impurities capable of absorbing sunlight. This explains, in most cases, the instability of polymers which, according to their chemical constitution, should be resistant to solar radiation.The probability that an absorbed photon induces a chemical change in a molecule depends on the importance of competing photophysical processes (radiationless transitions, luminescence, see Scheme 1) following the absorption act.

Contrary to UV and visible light, high energy radiation is absorbed non-specifically. i.e., there are no chromophores for γ-rays, fast electrons etc.. The absorption of high energy electromagnetic radiation and of fast particles by matter occurs via interactions with both the nuclei of the atoms and the clouds of electrons surrounding them. Interactions with atomic nuclei can be neglected if the energy of the photon or the particle is lower than about 10 MeV and if the irradiated material consists only of light nuclei (C,O,H,N,S,P), which applies to organic polymers.

Scheme 1. Important photophysical processes. (The superscripts 1,2 and 3 denote singlet, doublet and triplet states; asterisks denote excited states.)

Photon Absorption:

$^{1}M + h\nu \longrightarrow {}^{1}M^{*}$		(1st excited singlet state)
$^{1}M + h\nu \longrightarrow {}^{1}M^{**}$		(higher excited singlet state)
$^{1}M + h\nu \longrightarrow {}^{2}M^{+\cdot} + e^{-}$		(photoionization)

Radiationless Transitions:

$^{1}M^{*} \longrightarrow {}^{1}M$	+ energy	(internal conversion)
$^{1}M^{*} \longrightarrow {}^{3}M$	+ energy	(intersystem crossing)
$^{1}M^{**} \longrightarrow {}^{1}M^{*}$	+ energy	(internal conversion)
$^{3}M^{**} \longrightarrow {}^{3}M$	+ energy	(internal conversion)

Radiative Transitions (Luminescence):

$^{1}M^{*} \longrightarrow {}^{1}M$	$+ h\nu$	(fluorescence)
$^{3}M^{*} \longrightarrow {}^{1}M$	$+ h\nu$	(phosphorescence)

Regarding the interaction of γ- or X-rays with matter, three processes are to be considered: photoelectric effect, Compton effect and pair formation. The relative importance of these processes depends on the photon energy and the atomic number of the nuclei as well as on the electron density of the irradiated system. In all three cases secondary electrons are ejected from the atoms involved, most of them possessing kinetic energies sufficiently large to induce additional ionizations or electronic excitations in surrounding molecules:

$$M \xrightarrow{\text{high energy photon}} \underset{\text{radical cation}}{M^{+\cdot}} + \underset{\text{secondary electron}}{e^{-}_{kin}} \qquad (2)$$

$$M \xrightarrow{e^{-}_{kin}} M^{+\cdot} + e^{-}_{kin} \qquad (3a)$$

$$M \xrightarrow{e^{-}_{kin}} M^{*} \qquad (3b)$$

electronically excited molecule

Reaction (2) describes a primary ionization event and reactions (3a) and (3b) represent the interaction of secondary electrons with intact molecules. Notably, reactions (3a) and (3b) also apply for the absorption of corpuscular radiation, i.e. fast electrons and ions (protons, α-particle etc.) as far as the fraction of energy lost by inelastic collisions is concerned. The ionization of a molecule M by a fast electron is described by reaction (4):

$$e^-_{fast} + M \longrightarrow M^{+\cdot} + e^-_{kin} + e^-_{fast} \qquad (4)$$

primary electron		radical cation	secondary electron	primary electron
(E_{kin})			(E''_{kin})	(E'_{kin})
				$(E'_{kin} < E_{kin})$

E''_{kin} of most secondary electrons is less than 100 eV. Therefore, a large fraction of energy is deposited in close proximity to the place of original ionization. On the other hand, a fast electron of e.g. E_{kin}= 1 MeV, passes many molecules without interaction. Consequently, energy absorption occurs rather heterogeneously. According to radiation chemists' terminology a "track" of a fast electron consists of a series of "spurs" having an average distance of several hundred nm. The distribution of spurs is subject to the linear energy transfer, LET, of a particle traversing matter, i.e. the amount of energy dissipated per unit path length , - dE/dx. Concerning interactions of particles of energy E_{kin} and mass M with electron clouds of atoms of atomic number Z the LET is given by the Bethe equation:

$$-\frac{dE}{dx} = \frac{2\pi e^4 z^2 MZN}{E_{kin} m} \ln \frac{4E_{kin} m}{I_m M} \qquad (5)$$

e: elementary charge, m: mass of electron, z: number of charges per particle, N: number of atoms/cm^3, I_m: average ionization potential. At a given particle energy E_{kin},- dE/dx increases according to eq.(5) if Z and z increase.It must be pointed out that the average distance between spurs is reduced if the LET increases. Actually, overlapping of spurs leads to almost "cylindrical" tracks "filled" with ionizations and excitations. This applies to the absorption of particle radiations (protons, α-particles etc.). Pronounced differences in initial local concentrations of reactive intermediates are the basis of theories explaining the differences in XL and CS yields found on irradiating polymers with radiations of low and high LET.

With respect to chemical alterations caused by the absorption of high energy radiation in organic polymers, the elementary reactions listed in Scheme 2 have to be considered.
The formation of a stable product in a simple decay process as described by reaction (e) in Scheme 2 is a very rare event. The overwhelming portion of final products is formed via radical-radical or ion-ion (neutralization) processes.

Scheme 2. Interaction of high energy radiation with organic compounds. Generation and decay of intermediates.

Primary radiolytic act	$M \dashrightarrow M^{+\cdot} + e^-_{kin}$	(a)
Capture of thermalized electron by positive ion	$M^{+\cdot} + e^-_{th} \dashrightarrow M^*$	(b)
Capture of thermalized electron by neutral molecule	$M + e^-_{th} \dashrightarrow M^{-\cdot}$	(c)
Bond breaking reactions:		
Decomposition of excited molecule (radical formation)	$M^* \dashrightarrow R^\cdot_1 + R^\cdot_2$	(d)
Decomposition of excited molecule (formation of stable products)	$M^* \dashrightarrow A + B$	(e)
Decomposition of positive radical ion	$M^{+\cdot} \dashrightarrow R^\cdot + C^+$	(f)
Decomposition of negative radical ion	$M^{-\cdot} \dashrightarrow R^\cdot + D^-$	(g)
Side reactions:		
Charge transfer	$M^{+\cdot} + S \dashrightarrow M + S^{+\cdot}$	(h)
	$M^{-\cdot} + S \dashrightarrow M + S^{-\cdot}$	(i)
Energy transfer	$M^* + S \dashrightarrow M + S^*$	(j)

2. General Aspects Concerning XL and CS in Linear Polymers

2.1. RANDOM AND SPECIFIC SITE ATTACKS

Regarding homopolymers consisting of identical structural repeating units, the initiation of XL and CS under the influence of UV light or high energy radiation can be considered as random processes. This concerns the fact that all repeating units are equally likely to be affected during radiation absorption. A typical example of such a random process is the photolytical CS of ketone polymers, e.g. that of poly(phenylvinyl ketone) by a Norrish type II process:

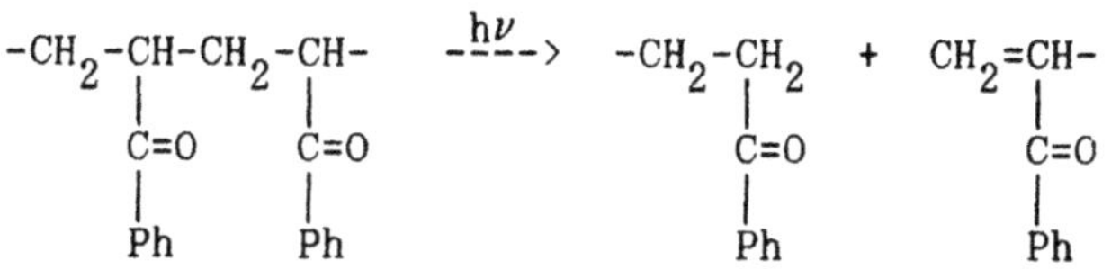

(6)

Upon irradiation with light of wavelength 300-370 nm, the carbonyl groups act as chromophores and, provided the polymer sample is thin enough to permit homogeneous distribution of absorption acts, CS occurs at random.

The occurrence of non-random processes can be visualized, for example, in the case of block copolymers of the type $-(A)_n-(B)-(A)_n,-$ consisting of long blocks of repeating units A and short sequences of repeating units B, with the latter being attacked exclusively. Such a situation applies, however, straightaway, only to photochemical initiation, where energy absorption is limited to chromophores. As far as high energy radiation is concerned principally all repeating units can be involved in absorption processes. The ensuing intermediates, e.g. free radicals might vary, however, in reactivity with the result that XL and/or CS are not distributed statistically over the chains of the block copolymer.

It should be pointed out that randomness of absorption events does not necessarily imply randomness of formation of reaction products. In heterogeneous systems, such as in partially crystalline polymers, products can be distributed also heterogeneously, if the product yields depend on the physical state of the system at the time of irradiation. This applies, e.g. to the γ-irradiation of polyethylene, where crosslinks are formed preferentially in the amorphous regions.

2.2. DETECTION OF XL AND CS

Both processes XL and CS alter the average molar mass (MM), and often also the molar mass distribution (MMD) of polymers. Therefore, all means of MM determination are applicable to evidence XL and CS provided the polymer sample is soluble in an appropriate solvent. In this case, average molar mass determinations can be carried out with the light scattering method which yields the weight average molar mass, M_w, and with osmotic methods (vapor pressure, membrane osmometry) which yield the number average molar mass, M_n. MMD's are determined by gel permeation chromatography which affords, however, calibration by an absolute molar mass determination method.

2.3. SIMULTANEOUS XL AND CS

Apart from a few cases, where main-chain scission occurs exclusively, linear polymers undergo both XL and CS upon exposure to UV light or high energy radiation. Then, the dependence of M_n and M_w on the absorbed dose D is given by the following equations:

PHOTOLYSIS:

$$\frac{M_{n,o}}{M_{n,D}} = 1 + (\phi(S)-\phi(X))\frac{D}{N_A} \tag{7}$$

$$\frac{M_{w,o}}{M_{w,D}} = 1 + (\phi(S)/2 - 2\phi(X))\frac{D}{N_A} \tag{8}$$

D: absorbed dose in photons/g; N_A: Avogadro's number,

$\phi(S)$ and $\phi(X)$: quantum yields, i.e. number of scissions and crosslinks, respectively, per photon

RADIOLYSIS:

$$\frac{M_{n,o}}{M_{n,D}} = 1 + (G(S)-G(X))\frac{D}{100\ N_A} \tag{9}$$

$$\frac{M_{w,o}}{M_{w,D}} = 1 + (G(S)/2 - 2G(X))\frac{D}{100\ N_A} \tag{10}$$

D: absorbed dose in eV/g; N_A: Avogadro's number, G(S) and G(X): number of scissions and crosslinks, respectively, per 100 eV.

Whereas eqs. (7) and (9) hold for any initial MMD, eq.(8) and (10) hold only for an initially random MMD. From eqs.(8) and (10) it can be seen, that M_w does not change if $G(S) = 4G(X)$ or $\phi(S) = 4\phi(X)$. If $G(S) > 4G(X)$, or $\phi(S) > 4\phi(X)$, the polymer degrades predominantly, i.e. M_w decreases with increasing absorbed dose. On the other hand, a three-dimensional, insoluble network is formed due to predominant

crosslinking, if $G(S) < 4G(X)$, or $\phi(S) < 4\phi(X)$.
$\phi(S)$ and $\phi(X)$, and $G(S)$ and $G(X)$ can be obtained separately if M_n and M_w are measured as a function of the absorbed dose. From eqs.(7) and (8) it results:

$$\frac{\phi(S)}{\phi(X)} = \frac{4k_n - k_w}{k_n - k_w} \tag{11}$$

with $k_n = \phi(S) - \phi(X)$ and $k_w = \phi(S)/2 - 2\phi(X)$.
Analogously, $G(S)$ and $G(X)$ can be obtained with the aid of eq.(12), derived from eqs. (9) and (10) :

$$\frac{G(S)}{G(X)} = \frac{4k'_n - k'_w}{k'_n - k'_w} \tag{12}$$

with $k'_n = G(S) - G(X)$ and $k'_w = G(S)/2 - 2G(X)$.
If crosslinking predominates, the polymer becomes insoluble at absorbed doses exceeding the gel dose, D_{gel}. In this case, $\phi(S)$ and $\phi(X)$, and $G(S)$ and $G(X)$ can be determined with the aid of the Charlesby-Pinner method according to eq.(13) or (14):

PHOTOLYSIS:

$$s + s^{1/2} = \frac{\phi(S)}{2\phi(x)} + \frac{N_A}{M_w \phi(X) D} \tag{13}$$

RADIOLYSIS:

$$s + s^{1/2} = \frac{G(S)}{2G(X)} + \frac{100\ N_A}{M_w G(X) D} \tag{14}$$

Here, s denotes the soluble fraction of the irradiated polymer sample. Eqs.(13) and (14) hold for simultaneous random XL and CS and for a random initial MMD.

2.4. MECHANISMS

Generally, alterations in the molar mass of polymers induced by UV light are easier to understand than those induced by high energy radiation. In the photolysis of polymers mostly only one kind of chromophore giving rise to a certain reaction mechanism is involved whereas high energy radiation generates simultaneously various kinds of intermediates which initiate different reaction routes. Principally, reaction mechanisms are quite specific and depend strongly on the chemical nature of the polymer. Some generalizations can be made, however, with respect to polymers containing carbonyl groups. Reaction (6) describes the light-induced chain cleavage of poly(phenyl vinyl ketone). In this case the excited carbonyl group abstracts a hydrogen from a carbon in γ-position. The biradical generated this way is short-lived and rearranges to the final products as described in reaction (6). Excitation of carbonyl groups might also induce α-scission, a process commonly referred to as Norrish type I reaction. Provided the carbonyl group is located in the chain, main-chain scission results from this reaction:

$$-CH_2-\underset{\underset{O}{\|}}{C}-CH_2- \xrightarrow{h\nu} -\dot{C}H_2 \quad + \quad {}^{\bullet}\underset{\underset{O}{\|}}{C}-CH_2- \qquad (15)$$

If the carbonyl groups are located in side chains, such as in the cases of poly(methyl acrylate), PMA, and poly(methyl methacrylate), PMMA, side groups are split off and lateral macroradicals are formed by the Norrish type I reaction. This was inferred from the fact that, upon UV irradiation, in both cases the same low molar mass products, especially $HCOOCH_3$, CH_3OH, CO_2 and CO are formed. That PMMA undergoes predominantly main-chain scission, whereas PMA crosslinks predominantly, is explained in terms of differences in the stability of the macroradicals generated by side group cleavage: $-CH_2-\dot{C}H-CH_2-$ is more stable than $-CH_2\dot{C}(CH_3)-CH_2-$. The mechanism is illustrated in Scheme 3.

Scheme 3. Photolysis of poly(methyl acrylate) and poly(methyl methacrylate)

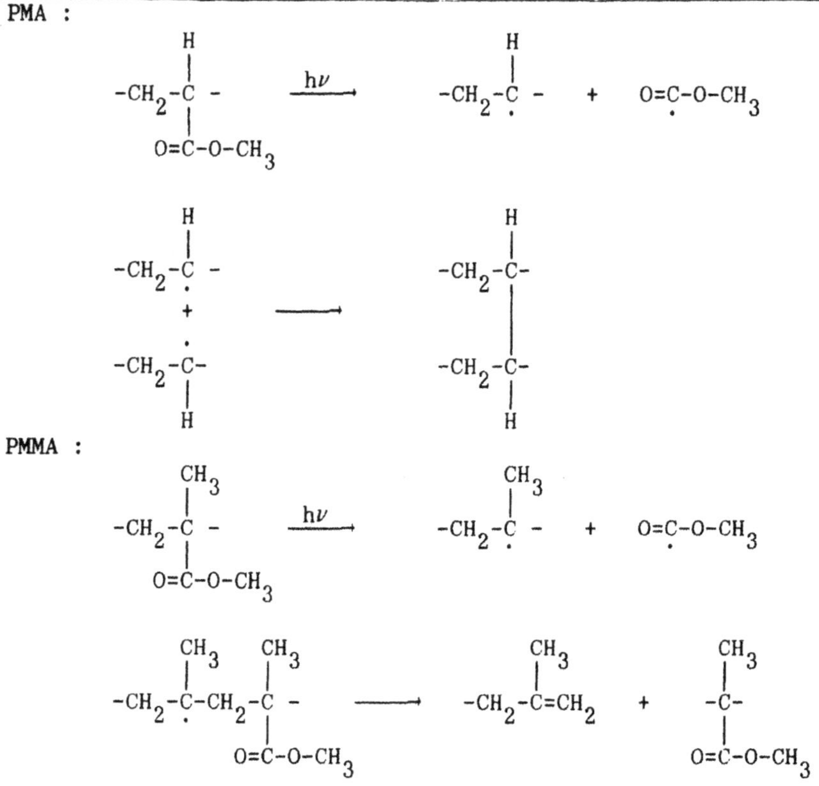

An example applying to crosslinking refers to the photolysis of poly-p-chloromethylstyrene, which was investigated in our laboratory (Zhu and Schnabel (1988)). In this case, a major light-induced reaction is the scission of the C-Cl bond in chloromethyl groups. The resulting chlorine radicals are reactive and abstract to a great extent hydrogens from carbons in the main chain. Macroradicals produced by both reactions form crosslinks by combination. This mechanism is illustrated in Scheme 4.

Scheme 4. Photochemical crosslinking of poly-4-chloromethylstyrene

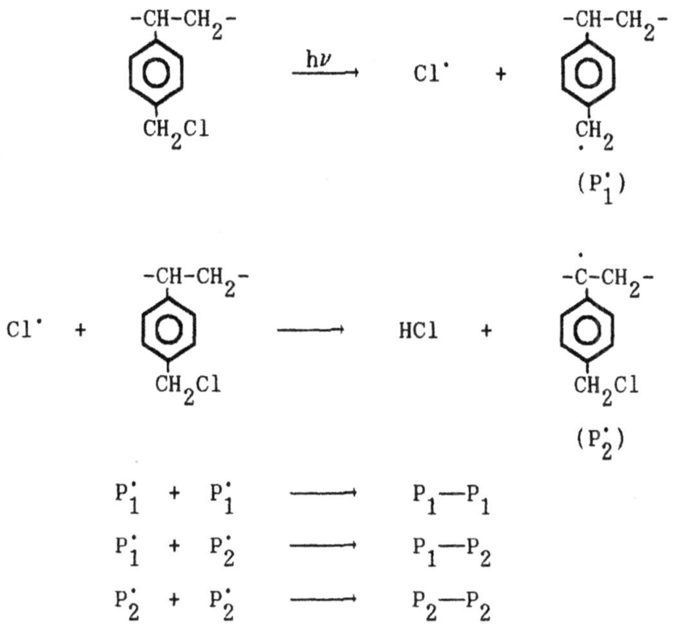

In Scheme 5 a mechanism concerning the radiolysis of poly(dimethyl itaconate) is proposed. The γ-ray - and electron beam-induced main-chain degradation of this polymer was recently studied in our laboratory (Popovic and Schnabel (1988)).

Pulse radiolysis experiments in conjunction with light scattering measurements showed that, in solution, main-chain degradation of poly(dimethyl itaconate) occurs in two modes. Only the fast mode was inhibited by N_2O, an effective electron scavenger. The slow mode, on the other hand, was inhibited by typical free radical scavengers such as O_2 and thiols (RSH). It was, therefore, concluded that the fast mode is due to the decomposition of highly excited states (P^{**}) which are formed by the reaction of thermalized electrons with radical cations. The slow mode, on the other hand, reflects the decomposition of (normally) excited states, probably excited keto groups according to a Norrish type I mechanism.

Scheme 5. γ-ray and electron beam radiolysis of poly(dimethyl itaconate)

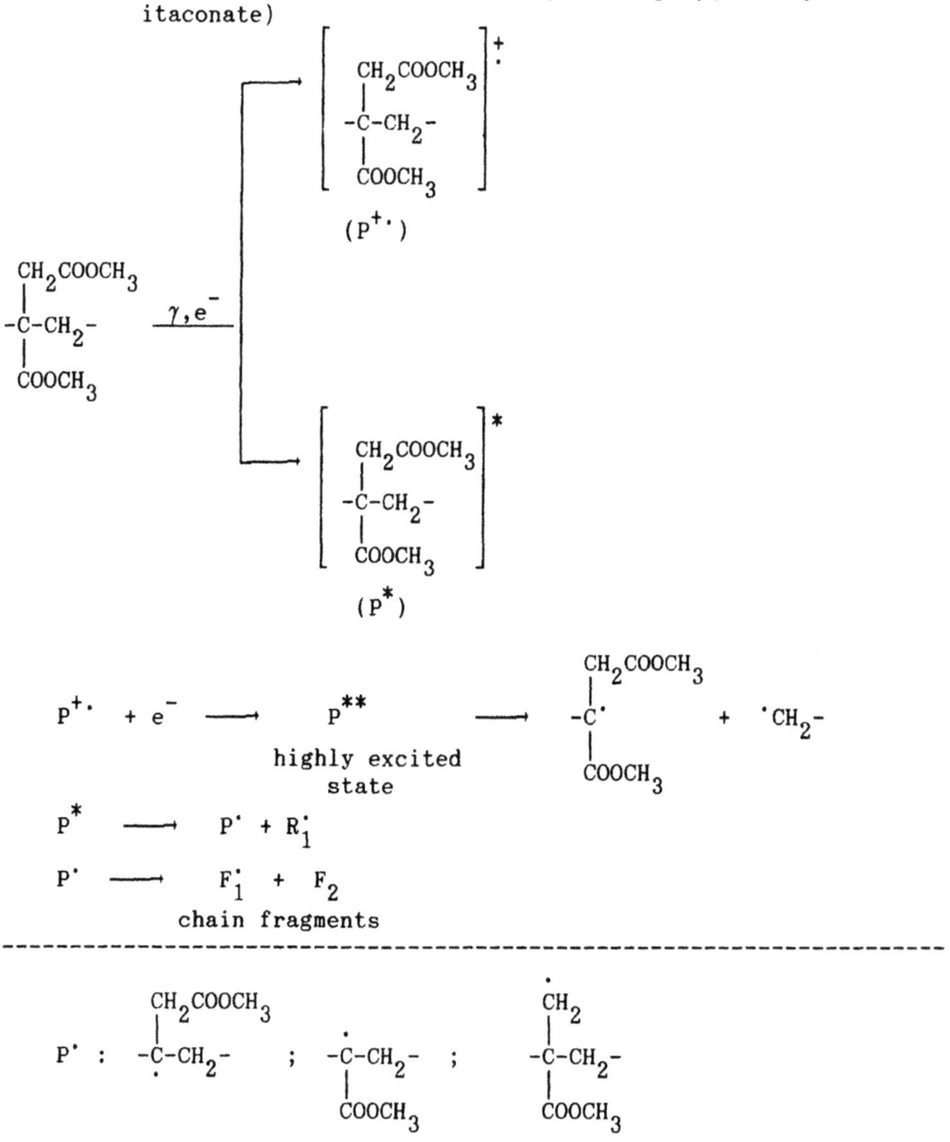

3. ION BEAM-INDUCED RADIATION EFFECTS IN LINEAR POLYMERS

It has been pointed out before (vide ante section 1.3) that, on passing through matter, heavy ions generate almost cylindrical tracks filled with intermediates, i.e. ions, excited molecules, free radicals. Because of the high local initial concentration of intermediates it is feasible to assume that radiation effects should differ qualitatively and quantitatively if polymers are subjected to radiations of different LET (-dE/dx). Actually, LET effects have been detected during the last decades. For example, G(S) was found to decrease with increasing LET in the case of poly(methyl methacrylate) (Zverev et al.(1969), Schnabel et al. (1984)) and G(X) was reported to be independent of LET in the cases of poly(vinyl chloride) (Makhlis (1975)) and polyethylene (Shinohara et al. (1958); Lavrentovich et al. (1969)). Polystyrene was reported to be crosslinked more efficiently by heavy ion radiation than by sparsely ionizing radiation (Parkinson et al. (1965); Zverev et al. (1969)).

Regarding the determination of G-values for high LET radiation various obstacles have to be overcome:
1) It is rather difficult to irradiate polymer samples homogeneously with heavy ion beams because of the small beam diameter (only a few mm). Moreover, the range of heavy particles is generally very short.
2) The LET is a function of the particle energy. Fig.2 shows a graph depicting -dE/dx= f(E) for α-particles.

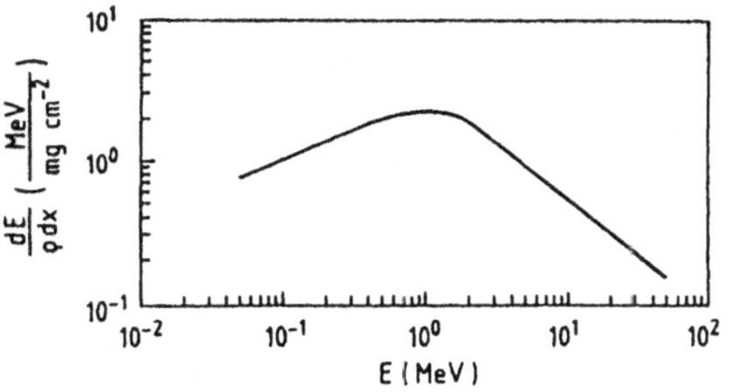

Fig.2. Stopping power - dE/dx vs. particle energy E. Absorption of α-particles in poly(methyl methacrylate)

Notably, -dE/dx increases and passes through a maximum with increasing E.
In a theoretical treatment by Chatterjee and Magee (1987) the track structure is characterized by two parameters, the "core radius" r_c and the" penumbra radius" r_p with $r_p > r_c$. $r_p - r_c$ is attributed to the volume surrounding the inner part of the track in which secondary electrons of relatively high energy, originating from the inner part, dissipate their energy. The inner part of the track is characterized by r_c. As has been outlined by Chatterjee and Magee (1987), the track structure varies with E. A typical calculation yielded that r_c and r_p vary with E in the following way: when E/nucleon decreases from 14 to 0.01 MeV/u, r_c decreases by at least one order of magnitude and r_p by at least three

orders of magnitude. Consequently, the local density of primary acts and the local density of reactive intermediates should differ significantly, if identical particles differing in E but not in dE/dx are compared. With respect to Fig.2 this implies that differences in the nature of chemical products and/or differences in product yields might be observable if a polymer is irradiated with the same particle radiation (differing in E_o) at equal stopping power (corresponding to points at the curve on both sides of the maximum). In praxi, heavy particles are totally absorbed in the polymer specimen and the track structure will be subject to alterations up to the track end. Therefore, track structure effects can only be evidenced experimentally, if the thickness of the polymer film traversed by the ion beam is much smaller than the particle range. Up to now experiments covering the energy range on both sides of the maximum of the curve in Fig.2 have not been carried out systematically.
3) A very high fraction of the absorbed dose is converted into heat (more than 90 %), i.e., only a small fraction of the particle energy serves to initiate chemical reactions. This implies, that the local temperature can increases drastically, if the short range of the particles is taken into account. Therefore, due to the low heat conductivity of organic polymers, the sample temperature is not easily controllable during irradiation.

In the following some results obtained recently in the author's laboratory are reported (Klaumünzer and Schnabel (1988)). Thin films of three polymers (d: 20-35 μm) were irradiated with 275 MeV $^{20}_{10}Ne^{7+}$ and 180 MeV $^{40}_{18}Ar^{8+}$ ions: poly(2-chloroethyl methacrylate),PMCMA, polymethacrylonitrile, PMCN, and polystyrene, PSt. The influence of temperature increase on the radiation chemical behavior could be studied best with PMCMA. This polymer crosslinks when heated in vacuo to temperatures above 220°C. Upon irradiating PMCMA with 180 MeV Ar ions at absorbed doses exceeding 1.2×10^{21} eV/g an insoluble gel was formed indicating the occurrence of intermolecular crosslinking. However, upon irradiating PMCMA films coated on one side with a 0.25 μm thick gold layer, the films remained soluble and the average molar mass decreased with increasing absorbed dose. Obviously, the gold layer provided for a fast transport of heat. The scission yield could not be determined in this case because a GPC calibration curve was not available for this polymer. - G(S) was determined, however, in the case of PMCN. Values are presented in Table 3:

Table 3. Main-chain scission of polymethacrylonitrile induced by radiation of different stopping power (LET).

Radiation	-dE/dx (MeV/μm)	G(S)	References
^{60}Co-γ-rays	$2x10^{-4}$	3.3	Helbert et al.(1980)
X-rays (λ≈ 1 nm)	$2x10^{-4}$	3.6	Sotobayashi et al. (1984)
$^{20}_{10}Ne^{7+}$ (E_o= 275 MeV)	0.43	0.5	Klaumünzer, Schnabel (1988)
$^{40}_{18}Ar^{8+}$ (E_o= 180 MeV)	2.5	0.4	Klaumünzer, Schnabel (1988)

Table 3 shows that G(S) decreases with increasing stopping power.

Notably, in the case of PSt the 100 eV yield of crosslinking was found to be almost independent of LET. This can be seen from Table 4.

Table 4. Crosslinking of polystyrene induced by radiations of different stopping power (LET)

Radiation	-dE/dx (MeV/μm)	G(X)	References
^{60}Co-γ-rays	$2x10^{-4}$	0.05	O'Donnell et al.(1979)
$^{20}_{10}Ne^{7+}$ (E_o= 275 MeV)	0.37	0.05	Klaumünzer, Schnabel (1988)
$^{40}_{18}Ar^{8+}$ (E_o= 180 MeV)	2.26	0.03	Klaumünzer, Schnabel (1988)

In this connection it is notable that Calcagno et al.(1987) reported G(X)= 0.3 for irradiation of PSt with 100 keV He, 200 keV Ne and 400 keV Ar ions covering the LET range from 0.21 MeV/μm to 0.85 MeV/μm. It should be pointed out that the total stopping power, dE/dx, is 0.37 MeV/μm both for 275 MeV and 200 keV neon ions and that, therefore, the differences in G(X) found by the two groups could reflect a strong track structure effect, as discussed before. A strong dependence of

G(X) on E was also reported recently by Aoki et al. (1988) for the irradiation of PSt with H^+ and He^+ ions. Also these authors interpret their results partly on the basis of a track structure effect. Actually, the beam intensity in the experiments with ions of relatively low energy was very high. Therefore, the possibility of a strong temperature increase in the polystyrene films during irradiation cannot be excluded. One should be careful, therefore, in arriving at final conclusions with respect to the track structure effect in this case.

4. X-RAY-INDUCED RADIATION EFFECTS IN LINEAR POLYMERS

A fascinating application regarding XL and CS applies to lithographic techniques for the production of microchips. Actually, polymers play a rather clandestine role here, since the final product does not contain polymer. Polymers are used, however, for protection of part of the wafer during processing. One of the important requirements on the polymer applies to (plasma) etch resistance , i.e. the polymer must be resistant against etching agents that are used to pattern the surface of the wafer. Therefore, appropriate polymers are denoted as "polymer resists". Lithographic techniques have to be rather sophisticated insofar as fine line structures of line width $\lesssim$ 0.3 μm must be generated in thin polymer films in order to produce 4 or 16 Mbit chips. To obtain fine line patterns of these dimensions, even deep UV radiation is insufficient. Quite appropriate are, however, electron beam radiation, generated by accelerators, and soft X-rays, provided by electron synchrotrons. For industrial large scale production only X-rays are applicable. In this case silicon wafers coated with a thin polymer film , about 1 μm thick, are irradiated through a finely structured mask. This way, the mask pattern is replicated in the polymer film. At present, liquid processing serves to develop the latent fine line pattern in the film.

In principle, positively working polymer resists are distinguished from negatively working ones depending on whether the exposed areas of the polymer film are rendered in the developer more soluble or less soluble than the unexposed areas. For many resist systems investigated during the last years crosslinking or main-chain scission were the essential radiation-induced chemical reactions giving rise to a decrease or increase in solubility, respectively. A major problem regarding the selection of appropriate polymers concerns the difficulty to combine high radiation sensitivity with high (plasma) etch resistance. One of the important criteria concerning the radiation sensitivity applies, of course, to the radiation chemical yields G(S) and G(X).

Investigations carried out in our laboratory with a positively working resist, polymethacrylonitrile , PMCN,

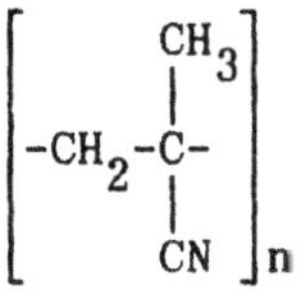

yielded that, apart from a high G(S) value, also the selection of appropriate development conditions is of great importance with respect to an improvement in the radiation sensitivity (Schlegel and Schnabel (1988)). The latter is defined here as the minimum exposure dose necessary to achieve full development of the latent line pattern in the polymer film. G(S)= 3.3 was found both for irradiations with ^{60}Co-γ-rays and soft X-rays ($\lambda_{inc} \approx 0.1$-$1.0$ nm), respectively.
In the case of positively working resists, the molar mass of the polymer should be reduced significantly, say, by a factor of 5, to attain a sufficiently large difference in solubility between irradiated and unirradiated areas of the polymer film. From eq.(10) it follows that the absorbed dose D_5, resulting in $(M_{w,o}/M_{w,D}) = 5$, is

$$D_5 = \frac{800\ N_A}{G(S)\ M_{w.o}} \qquad (16)$$

According to eq.(16) D_5 is inversely proportional to the initial molar mass $M_{w,o}$ which implies that $M_{w,o}$ should be as high as the limitations, set by the processibility in spin-coating, allow.
In the case of PMCN it turned out, however, that the initial molar mass did not affect the radiation sensitivity to the extent expected from eq.(16). Much more important than the employment of a polymer sample of high initial molar mass was the selection of solvents capable of dissolving the irradiated polymer portions without swelling the unirradiated ones
Regarding the dissolution of amorphous polymers at temperatures below the glass transition ($T<T_g$) the following steps have to be discriminated:
(a) Solvation: penetration of solvent molecules into the pores of the rigid matrix.
(b) Relaxation: transition of the polymer from the glassy to the equilibrated state
(c) Reptation: disentanglement and diffusion of macromolecules towards regions of low polymer concentration
(d) Convection: transport of the macromolecules by the flowing solvent.
In order to generate fine line patterns of high quality the insoluble portions should remain in the glassy state during the development. The contrast between exposed and unexposed areas is determined by the ratio of the solvation rates $v_{s,D}/v_{s,o}$. This ratio is the higher the weaker

the interaction is between polymer and solvent. Here, the Hansen parameter of solvents served as a measure for the strength of interaction during the selection of appropriate liquids. This way, cyclohexanone, acetophenone, benzonitrile and cyclopentanone were selected and it was found that the surface of PMCN films did not swell during development with these solvents. Solvation was the rate determining step. This explains, why, in the case of PMCN, the radiation sensitivity is only to some extent affected by the radiation-induced decrease in average molar mass. Actually, it is assumed that differences in the dissolution rates referring to irradiated and unirradiated areas are essentially due to an increase in the free volume of the glassy polymer which could be a consequence of the formation of micromolecular gaseous degradation products. This assumption was substantiated by the fact that the dissolution rate was proportional to the exposure dose.

Although high quality fine line patterns could be replicated with the aid of the solvents mentioned above, obstacles still existed with respect to rather long times of development combined with a partial dissolution of unexposed areas of the polymer film. These difficulties were overcome by selective pre-swelling of the unirradiated areas. For this purpose liquids were used that interact only very weakly with PMCM, for example methylpropylketone or diethylketone. These liquids are incapable of dissolving PMCN but are capable of penetrating into the film and transforming it thereby from the glassy to the relaxed state. Consequently, the subsequent development is significantly accelerated. In typical experiments, in which high quality fine line patterns of 0.3 μm line were obtained at an exposure dose of about 200 mJ/cm^2, the pre-swollen polymer films were developed for 2 minutes in a mixture of cyclohexanone/o-xylene/water at a weight ratio of 65/34/1. From the results obtained in this work it can be concluded that, although the importance of a high yield of main-chain scission should not be neglected, the employment of proper development conditions is crucial for the generation of fine line patterns in polymers at relatively low exposure doses.

REFERENCES

Polymer Degradation

- B. Dolezel, " Die Beständigkeit von Kunststoffen und Gummi", Hanser, München (1978)
- W.Schnabel,"Polymer Degradation, Principles and Practical Applications", Hanser, München (1981)
- H.H.G.Jellinek, ed., Aspects of Degradation and Stabilization of Polymers", Elsevier, Amsterdam (1978)
- T.Kelen, "Polymer Degradation", Van Nostrand Reinhold Co., New York (1983)
- L.Reich and S.S.Stivala, "Elements of Polymer Degradation", McGraw-Hill, New York (1971)
- N.Grassie and G.Scott, "Polymer Degradation and Stabilization",Cambridge University Press, Cambridge (1985)

Photochemistry of Polymers

- W.Schnabel and J.Kiwi, Photodegradation", in H.H.G.Jellinek, ed., "Aspects of Degradation and Stabilization of Polymers", Elsevier,Amsterdam (1978)
- G.Geuskens, "Photo-degradation of Polymers", in C.H.Bamford and C.F.H.Tipper, eds., Elsevier (1975)
- J.F.McKellar and N.S.Allen, "Photochemistry of Man-Made Polymers", Applied Science Publ., London (1979)
- N.S.Allen, ed., "Developments in Polymer Photochemistry", Vol.1-3, Applied Science Publ., London (1980-1982)
- V.A.Shlyapintokh, "Photochemical Conversion and Stabilization of Polymers", Hanser, München (1984)
- J.Guillet, "Polymer Photophysics and Photochemistry" Cambridge University Press, Cambridge (1985)
- J..G.Kloosterboer, "Network Formation by Chain Crosslinking. Photopolymerization and its Applications in Electronics", Adv.Polym.Sci 84, 1 (1988)

Radiation Chemistry of Polymers

- W.Schnabel,"Degradation by High Energy Radiation", in H.H.G.Jellinek, ed., "Aspects of Degradation and Stabilization of Polymers", Elsevier, Amsterdam (1978)
- A.Makhlis, "Radiation Physics and Chemistry of Polymers", J.Wiley & Sons, New York (1975)
- M.Dole, ed., "The Radiation Chemistry of Macromolecules", Academic Press, New York (1972)
- A.Charlesby, "Atomic Radiation and Polymers", Pergamon Press, Oxford (1960)
- A.Chapiro, "Radiation Chemistry of Polymeric Systems", Interscience, New York (1962)

Reaction Mechanisms

- Q.Q.Zhu and W.Schnabel, Polym.Degrad.Stab. in print
- I.Popovic and W.Schnabel, Eur.Polym.J. submitted

LET Effects

- Y.Aoki, N.Kouchi, H.Shibata, S.Tagawa and Y.Tabata Nucl.Instr.Meth.Phys.Res. B33, 799 (1988)
- A. Chatterjee and J.L.Magee, in Farhataziz and M.A.J.Rodgers, eds., "Radiation Chemistry, Principles and Applications", VHC Publ., Weinheim (1987), Chapters 1,5 and 6
- L.Calcagno, G.Foti, A.Licciardello and O.Puglisi, Appl.Phys.Lett. 5,907 (1987).
- N.Helbert, G.Iafrate, C.U.Pittman and J.H.Lai, Polym.Eng.Sci.20, 1077 (1980)
- F.A.Makhlis, "Radiation Physics and Chemistry of Polymers", J.Wiley, New York (1975)
-.J.H.O'Donnell, N.P.Rahman, C.A.Smith and D.J.Winzor, J.Polym.Sci., Polym.Chem.Ed. 17, 4081 (1979)
- W.W.Parkinson, C.D.Bopp and D.Binder, J.Phys.Chem. 69, 604 (1965)
- W.Schnabel, S.Klaumünzer, H.Sotobayashi, F.Asmussen and Y.Tabata,
- Macromolecules 17, 2108 (1984)
- W.Schnabel and S.Klaumünzer, Rad.Phys.Chem, in print
- K.Shinohara, A.Amemiya and A.Danno, J.Phys.Soc.Japan, 13, 604 (1958)
- H.Sotobayashi, F.Asmussen, Gunawan M.Angga, N.Abou-Samra, W.Schnabel and F.P.Wolf, BESSY, Anual Report, p.190 (1984)
- A.B.Zverev, Ya. Lavrentovich and A.M.Kabakhi, Khim.Vys.Energ. 3,453 (1969)

Lithography

- F.A.Vollenbroek, "Photoresist Systems for Microlithography", Adv.Polym.Sci. 84, 85 (1988)
- W.S.DeForest, "Photoresist, Material and Processes", McGraw-Hill, New York (1975)
- L.F.Thompson, C.G.Willson and M.J.Bowden, eds., "Introduction to Microlithography", ACS Symposium Series 219, Washington (1983)
- L.Schlegel and W.Schnabel, J.Vac.Sci.Techn.B 6, 82 (1988)
- L.Schlegel, " Zur Optimierung von Polymerresists für die Röntgenlithographie - Untersuchungen über Entwicklersysteme und Prozeßführung bei der Naßentwicklung", Dissertation, Technische Universität Berlin (1988)
- C.M.Hansen, "Three dimensional solubility paramater and solvent diffusion coefficient", Dänische Technische Presse, Kopenhagen (1967); J.Paint Technol. 39, 104 (1967); 39, 505 (1967)
- A.E.Novembre, L.M.Masakowski and M.A.Hartney, Proceed. Conf. on Photopolymers: Principles,Processes and Materials, Ellenville, October (1985)

MAIN-CHAIN SCISSION OF LINEAR POLYMERS AS STUDIED BY PULSE RADIOLYSIS AND FLASH PHOTOLYSIS

WOLFRAM SCHNABEL
HAHN-MEITNER-INSTITUT, BEREICH S
GLIENICKER STR.100
FEDERAL REPUBLIC OF GERMANY

ABSTRACT. This article demonstrates how main-chain degradation can serve to study motions of macromolecules and to obtain information on segment-segment interactions on the basis of light scattering (LS) measurements in conjunction with flash photolysis or pulse radiolysis. The LS detection method is a powerful tool to study kinetics and mechanisms of reactions of macromolecules in solution. Moreover, it can be used to study dynamic phenomena of polymer solutions. This article reports on measurements concerning the dynamics of synthetic and biological polymers. Preference is given here to problems concerning the lifetime of polymer-polymer interactions in dilute solutions. The basis for these studies are fast chemical reactions leading to main-chain scission. In certain cases the rate of the chemical reaction is much faster than the rate of consecutive dynamic processes involving the motion of polymer fragments and finally leading to the separation of the fragments. In these cases the rate of LS intensity change reflects the rate of fragment motion. Typical results applying to the lifetime of contact pairs formed by interacting repeating units of some synthetic linear polymers and to the lifetime of salt bridges formed between strands of biopolymers are reported.

1. Introduction

Mechanistic and kinetic studies concerning radiation-induced crosslinking and main-chain scission reactions in polymers in solution can be carried out conveniently with the aid of light scattering (LS) measurements [Beck et al. (1974); Schnabel (1979]. Appropriate reaction conditions conform to irradiation times of nanoseconds to microseconds, which are quite normal for conventional flash photolysis or pulse radiolysis set-ups, where a solution is irradiated with a short pulse of radiation. Changes in molar mass or molecular size of the macromolecules during and after the pulse can

O. Güven (ed.), Crosslinking and Scission in Polymers, 37–56.

be detected by a change in the intensity of light scattered by the solution. For this purpose a continuous beam of analyzing light is permitted to pass the solution. In our case the 514 nm line generated by an argon ion laser was found appropriate for most experiments.
According to the Debye equation (1) the light scattering intensity

$$\frac{Kc}{R_\theta} = \frac{1}{M_w} + \frac{16\,\pi^2 \langle s^2 \rangle}{3\,\lambda_o^2\,M_w} \sin^2(\theta/2) + 2A_2 \qquad (1)$$

R_θ (Rayleigh ratio), measured at an angle θ, is correlated to the weight average molar mass M_w, the mean square radius of gyration $\langle s^2 \rangle$, the polymer concentration c , the wave length of the incident analytical light and the 2nd virial coefficient A_2. $K = (2\pi^2 n_o^2 / N_A \lambda_o^4)$ $(dn/dc)^2$, where n_o is the refractive index, dn/dc the refractive index increment and N_A Avogadro's number.
According to eq.(1) R_θ is expected to increase or decrease with increasing or decreasing M_w, respectively. This applies especially to intermolecular crosslinking or main-chain scission, respectively. Notably, changes in R_θ are expected at constant M_w, moreover, if the radius of gyration is changed. Such changes are generated, e.g., by expansion or collapse of coil-shaped macromolecules. As can be concluded from eq.(1) coil expansion leads to a decrease in R_θ whereas coil shrinkage causes an increase in R_θ. Obviously, a variety of problems can be studied with the aid of light scattering measurements. This article is devoted to a few examples which all apply to motions of macromolecules in solutions [Beck et al. (1975a); Beck et al. (1975b); Denk et al. (1983); Dobrowolski et al. (1976); Dobrowolski, Schnabel (1979); Horie, Schnabel (1982, 1984); Lindenau et al. (1976a, 1976b); Lindenau et al. (1977); Tagawa, Schnabel (1979a, (1979b); Washino et al. (1983)]
Actually, in most cases referred to here advantage was taken from the fact that the light scattering intensity (LSI) is decreased in the case of main-chain scission. As a matter of fact, main-chain scission is a chemical reaction and one is tempted to assume that the rate of LSI change should reflect the rate of this chemical reaction. Actually this applies, however, only to those cases where rather long-lived intermediates are involved in main-chain scission, for example, to the case that a free radical site located at a macromolecule decays by main-chain rupture:

-------•---------- ---> -------• + ---------- (2)

More generally spoken, main-chain scission consists normally in a series of consecutive reactions: chemical steps involving the generation and decomposition of reactive sites, and physical steps involving the separation of fragments after bond cleavage by diffusion. Now, two extreme cases can be distinguished:
Case (a). The chemical steps are very fast, i.e. much faster than the physical steps. Consequently, the rate of LSI decrease corresponds to the rate of the "rate determining" physical steps.
Case (b). The physical steps are much faster than the chemical steps. Here, the rate of LSI decrease reflects the rate of one or several chemical reaction(s).
As has been pointed out (vide ante) this article only deals with dynamic aspects. Work devoted to applications of the LS detection method to studies concerning the elucidation of mechanisms and kinetics of chemical reactions of synthetic polymers and biopolymers will not be reviewed here. Instead the reader is referred to other recent reviews [Schnabel 1986, 1988)] and to the original papers [Beavan, Schnabel (1978); Beavan et al. (1978); Deeg et al. (1984, 1985, 1987); Gröllmann, Schnabel (1980); Kihara, Schnabel (1980) Rosiak, Schnabel (1984, 1986); Seki, Schnabel (1982); Washino, Schnabel (1982a, 1982b); Washino et al. (1984)].

2. Studies on the Dynamics of Macromolecules in Solution

From the previous section it can be concluded that the rate of motion of macromolecules can be measured if the chemical processes resulting in main-chain scission are much faster than fragment diffusion. Other prerequisites pertaining to the application of this method to dynamic studies are that the polymer does not crosslink, i.e. that it undergoes exclusively main-chain scission and that the refractive index increment, dn/dc, is sufficiently high. So far, only a few polymers fulfilling these conditions were found: polyisobutene (PIB), poly(phenylvinyl ketone) (PPVK) and poly(styrene sulfone) (PSS). Whereas PIB is readily degraded by high energy radiation (electron beam or γ-radiation) PPVK and PSS undergo main-chain rupture upon UV-irradiation as illustrated by reactions (3) to (5):

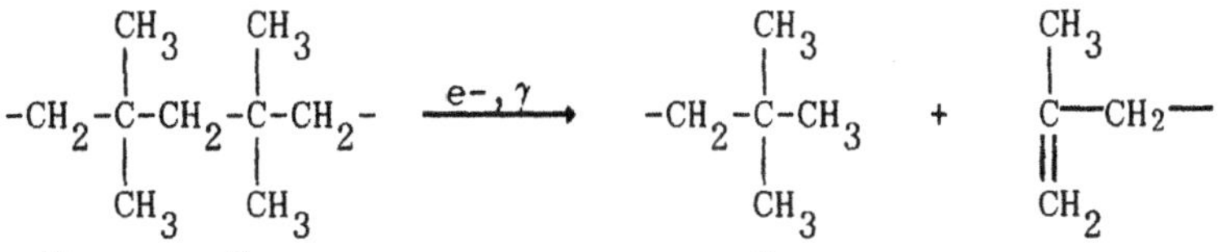

(3)

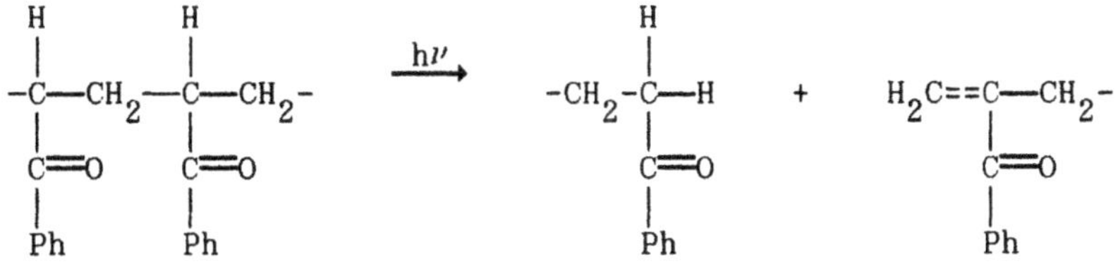

(4)

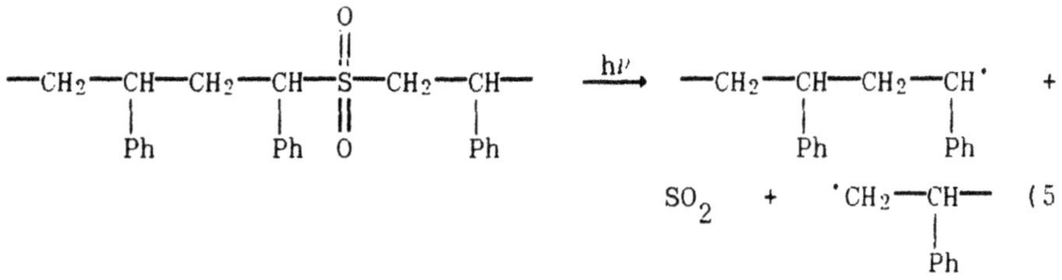

(5)

Poly(butene-1 sulfone), PBS, exhibited a behavior similar to that of PSS when main-chain scission was sensitized by benzene. In this case dioxane solutions of PSB containing a small amount of benzene were irradiated at λ_{inc}= 265 nm where PSB is transparent and main-chain scission was induced by energy transfer:

$$\text{Benzene} + h\nu \longrightarrow \text{Benzene}^* \quad (6)$$

$$\text{Benzene}^* + \text{PSB} \longrightarrow \text{Benzene} + \text{PSB}^* \quad (7)$$

$$\text{PSB}^* \longrightarrow \text{Fragments} \quad (8)$$

Fig.1 shows typical results obtained with PIB in hydrocarbon solutions. The kinetic traces shown in Fig.1(a) depict the decrease in LSI after a pulse of 16 MeV electrons was applied to the solutions. Fig.1(b) shows 1st order plots.

Evidence for the fact that the kinetic traces reflect the diffusion of fragments after main-chain rupture were obtained in the following ways: Cyclohexene which is known to inhibit main-chain rupture was added to solutions of cyclohexane. As can be seen from Table 1 the

Table 1. Main-chain scission of polyisobutene in cyclohexane upon electron beam irradiation (16 MeV electrons). Influence of added cyclohexene on the extent of degradation and the halflife of LSI decrease. [PIB]= 0.018 base mol/l. $M_{w,o}$= 2.1×10^6, $M_{n,o}$= 1.3×10^6. $D_{abs}\approx$ 250 Gy. [Beck et al. (1975)].

[Cyclohexene] (volume %)	$\frac{U_o - U_\infty}{U_\infty - U_L}$	$\tau_{1/2}$ (μs)
0	0.55	23 ± 1.5
1	0.26	24 ± 1.5
2	0.25	25 ± 1.5
5	0.16	26 ± 1.5
10	0.14	26 ± 1.5

extent of LSI decrease was significantly reduced whereas the half-life remained constant within the error limit upon increasing the cyclohexene concentration. Moreover, the halflife increased linearly with the microviscosity of the solution. This is demonstrated by Fig.2, where the halflife measured in different hydrocarbons is plotted vs. the viscosity of the solvent.

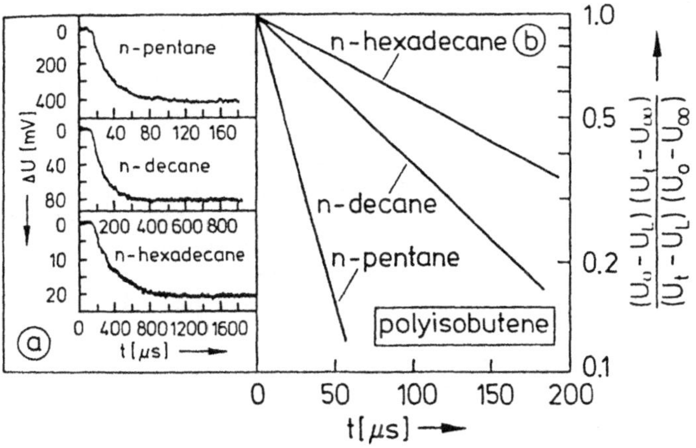

Fig.1. Main-chain degradation of polyisobutene in hydrocarbon solutions at room temperature. (a) Kinetic traces depicting the change in LSI in different solvents after irradiation with a pulse of 16 MeV electrons. (b) 1st order plots of the time dependence of the LSI decrease. U_o, U_t, U_∞ and U_L denote signal voltages being proportional to the LSI. Denotations of subscripts: (o) before the pulse, (t and ∞) at time t and a long time after the pulse, respectively, (L) solvent. $M_{w,o}$= $2.1x10^6$, $M_{n,o}$= $1.3x10^6$. [PIB]= $2.7x10^{-2}$ base mol/l. D_{abs}= 400 Gy/pulse. Pulse length: 2 μs. [Beck et al. (1975)].

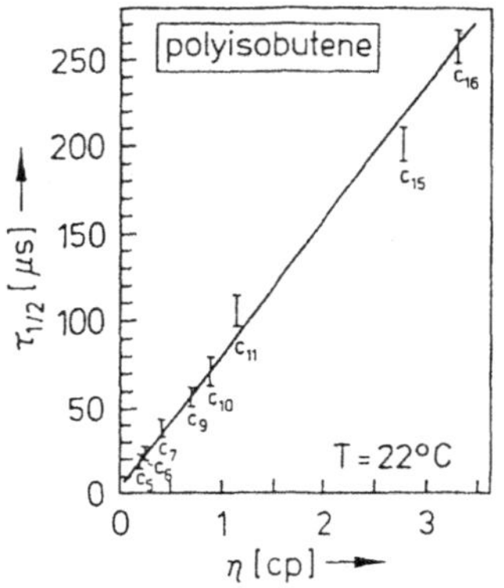

Fig.2. Main-chain degradation of polyisobutene in hydrocarbons (n-pentane to n-hexadecane) at room temperature. Dependence of the halflife of LSI decrease on the microviscosity (viscosity of the neat solvent). For experimental conditions: refer to legend to Fig.1. [Beck et al. (1975)].

In the case of PPVK main-chain scission is readily induced by UV irradiation. Bond scission follows the Norrish type II mechanism. A biradical is formed as an intermediate from an excited triplet state. The lifetimes of both triplet state and biradical are shorter than 100 ns if the reaction is carried out in solution at room temperature, for example in benzene solution. This was inferred from flash photolysis studies performed in conjunction with optical absorption measurements. As can be seen from Fig.3 the change in the LSI caused by main-chain scission occurs in the µs-range i.e. at a rate much higher than that of the chemical process indicating that the rate of LSI change reflects the rate of fragment diffusion.

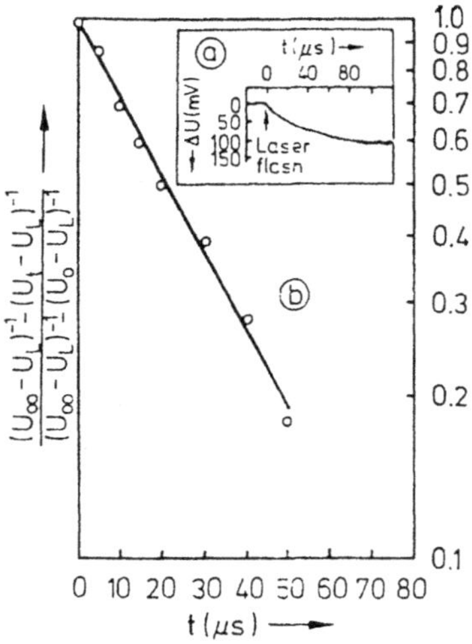

Fig.3. Main-chain degradation of PPVK in benzene solution at room temperature. [PPVK]= 3.8×10^{-3} base mol/l, $M_{w,o} = 4.6 \times 10^5$. $\lambda_{inc} = 347$ nm. Flash duration: 25 ns.
(a): kinetic trace depicting the decease in LSI
(b): 1st order plot
[Beck et al. (1974)]

3. Studies concerning intramolecular interactions

As far as quite dilute solutions of flexible linear macromolecules are concerned two kinds of physical interactions have to be considered: interactions of macromolecules with solvent molecules and interactions of structural repeating units belonging to the same macromolecule. At higher concentrations also interactions of repeating units of different macromolecules become important. The interactions of repeating units, which are denoted here as "contact pairs", endure only for limited periods. In the case of polymers consisting of flexible linear macromolecules contact pairs are statistically distributed along the chain. They are formed and decomposed perpetually, and a dynamic equilibrium between repeating units involved in contact pairs and those not involved exists. The average number of contact pairs per macromolecule is determined by the equilibrium constant which depends on the solvent power. It is relatively high in a "poor" solvent and low in a "good" solvent.
In the following it will be shown how the lifetime of contact pairs was determined with the aid of flash photolysis in conjunction with light scattering measurements.
During investigations on the degradation of polyisobutene in n-hexane solution it was found that the lifetime of LSI decrease soared upon the addition of a precipitant, in this case n-propanol. This can be seen from Fig.4, where $\tau_{1/2}$(LSI) is plotted vs. the volume fraction of the precipitant. These data seemed to indicate that the number of contact pairs increased upon addition of precipitant to the solution. Consequently, the rate of LSI change appeared to be determined by the lifetime of contact pairs. The same conclusion was arrived at by comparing the experimentally determined dependence of $\tau_{1/2}$ on the initial molar mass with that expected for free translational diffusion of the fragments. For the latter case it is expected [Beck et al. (1975); Schnabel (1979)] that the following proportionality holds:

$$\tau_{1/2} \propto M^{a} \qquad (9)$$

with a = 1.5-1.9.
However, for hydrocarbon solutions a = 0.34 was found, indicating that fragment separation is not a free translational diffusion process.
The model used to interpret the experimental results is illustrated in Scheme 1, where one can see that a flexible coiled macromolecule possessing several contact pairs is suddenly split into two fragments by a very rapidly occurring chemical reaction. After the chemical reaction the fragments are still held together by contact pairs which are acting as physical bonds. Due to their transitory

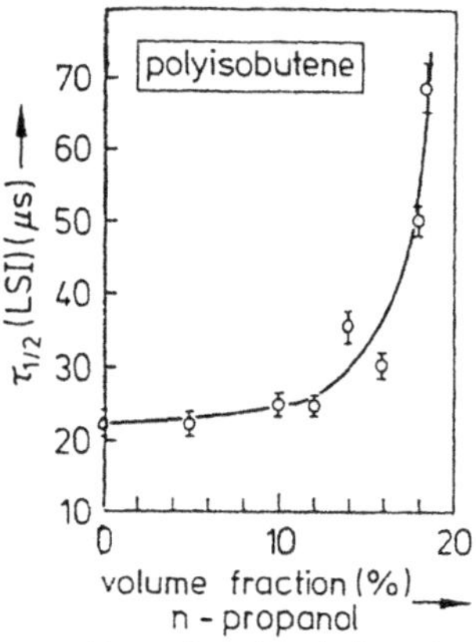

Fig.4. Degradation of polyisobutene in n-hexane/n-propanol mixtures at room temperature. The lifetime of LSI decrease vs. the volume fraction of n-propanol. For experimental conditions refer to legend to Fig.1.

nature the physical bonds are broken up after some time and the fragments become free to move independently. Detachment of contact pairs is the rate determining process if it is slower than the subsequent diffusion of the fragments. It appears that this applies whenever Z_{CP}, the average number of contact pairs per macromolecule, is high enough to ensure fragment attachment after main-chain scission, irrespective of the position of bond rupture at the chain. Therefore, the influence of contact pairs on the rate of chain fragment separation is expected to become strongly pronounced if the number of contact pairs per initial macromolecule is high, i.e exceeding significantly unity. Taking this into account, the increase in $\tau_{1/2}$ with increasing precipitant concentration shown in Fig.4 becomes intelligible. Obviously, Z_{CP} does not increase linearly with precipitant concentration, as indicated by the steep increase of $\tau_{1/2}$ at relatively high precipitant concentrations.

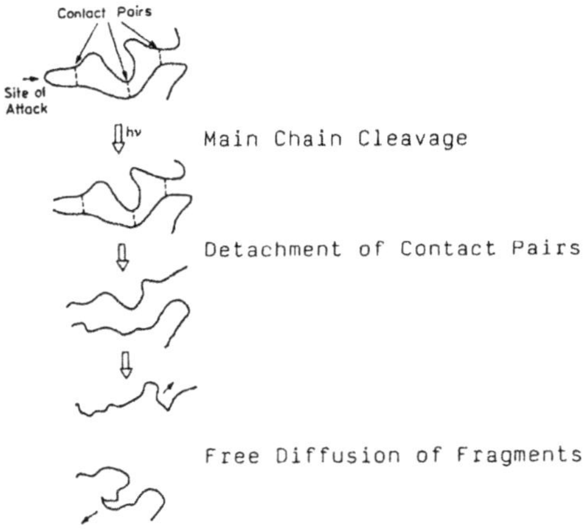

Scheme 1. Illustration of the detachment of a coiled macromolecule possessing several intramolecular contact pairs after bond rupture in the main chain

A systematic investigation on this topic was carried out with PPVK [Lindenau, Schnabel (1978), Tagawa, Schnabel (1979a), (1979b)]. In this case, Z_s, the number of scissions per initial macromolecule, could easily be varied by varying the intensity of the photolyzing light incident upon the polymer solution, because of the relatively high quantum yield of main-chain scission ((ϕ(S)≈ 0.2-0.3). The average fragment size decreases with increasing Z_s. Typical results are presented in Fig.5, where the halflife of LSI decrease is plotted vs. Z_s.

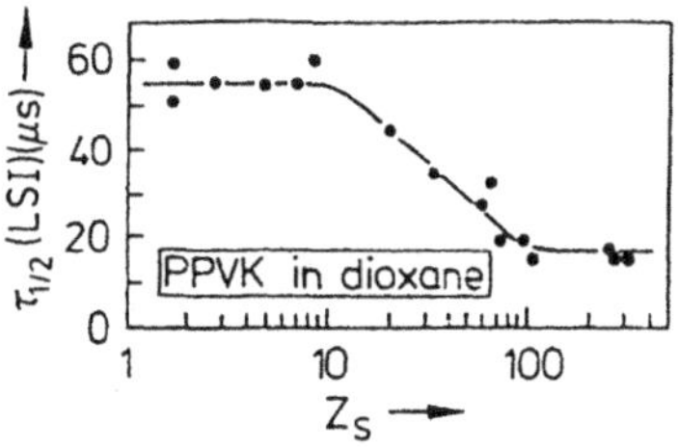

Fig.5. Main-chain degradation of poly(phenyl vinyl ketone), PPVK, in p-dioxane solution at room temperature upon irradiation with a 25 ns flash of 347 nm light. $\tau_{1/2}$, the halflife of LSI decrease vs. Z_s, the number of scissions per initial macromolecule. $M_{w,o}$= 1.5x10^6. [PPVK]= 3.8x10^{-3} base mol/l.[Tagawa, Schnabel (1979a)]

Notably, at low values of Z_s, corresponding to large fragments, $\tau_{1/2}$ is independent of Z_s. If Z_s exceeds a critical value, Z_s(crit), $\tau_{1/2}$ decreases with increasing Z_s corresponding to decreasing average fragment size. In other words, if the macromolecules are simultaneously split into many fragments (at high Z_s values) contact pairs have no or little influence on the rate of fragment separation. In this case the measured rate of LSI decrease reflects essentially free diffusion. On the other hand, if the macromolecules are only split into a few fragments, contact pairs exert a significant influence on fragment separation. At Z_s> Z_s(crit) translational diffusion becomes rate determining and at very high values of Z_s, i.e. at low fragment size, the rate of fragment detachment is constant, at least within the error limit of the measurements. On the basis of $\tau_{1/2}$ ≈ 55 μs for Z_s< Z_s(crit) (see Fig.5) the lifetime of contact pairs should be about 80 μs. This would apply to the case that decomposition of contact pairs solely determines fragment detachment. In reality, the situation is more complicated as can be seen from Fig. 6, where $\tau_{1/2}$ measured at various scattering angles, θ, and various initial molecular weights of the polymer, $M_{w,o}$, is plotted vs. Z_s.

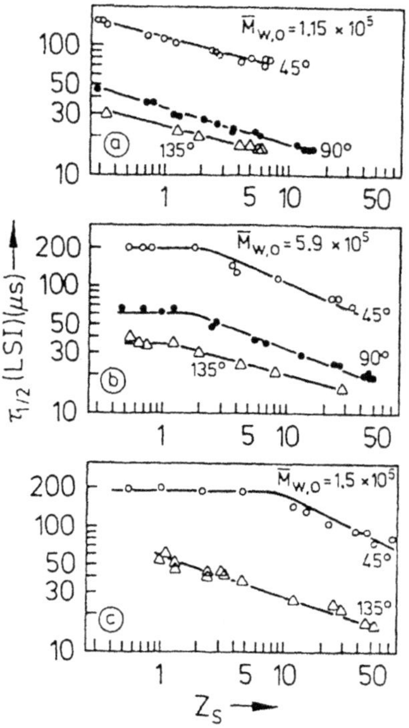

Fig.6
Main-chain degradation of poly(phenylvinyl ketone) in p-dioxane solution at room temperature upon irradiation with 25 ns flashes of 347 nm light. $\tau_{1/2}$(LSI) as a function of Z_s, the number of scissions per initial macromolecule, at various scattering angles. [PPVK]= 5×10^{-3} base mol/l. $M_{w,o}$= 1.15×10^5(a), 5.9×10^5 (b) and 1.5×10^6 (c). [Tagawa, Schnabel (1979b)]

Notably, at $M_{w,o}$= 1.15×10^5, $\tau_{1/2}$ decreased steadily with increasing Z_s indicating the absence of a significant influence of intramolecular interaction on the rate of fragment detachment in this case. With polymer samples of higher $M_{w,o}$(5.9×10^5 and 1.5×10^6) $\tau_{1/2}$ was found to be independent of Z_s at Z_s < Z_s(crit). At constant θ, Z_s(crit) increased with increasing $M_{w,o}$. Therefore, it was concluded that intramolecular interaction was determining the rate of fragment separation if the degree of polymerization exceeded about 3000. Another interesting feature regarding the data in Fig. 6 applies to the decrease of $\tau_{1/2}$ with increasing θ. This effect can be explained on the basis of the fact that main-chain ruptures are distributed statistically over all molecules and along the chains. Therefore, a statistical fragment size distribution is generated upon irradiation. At low scattering angles, the large fragments will contribute more to the light scattering intensity than the small ones, since large particles scatter more strongly in the forward direction than in the reverse direction. In other words at low angles one "sees" more of the large fragments than of the small fragments. On comparing the halflifes corresponding to the range $Z_s < Z_s(\text{crit})$ it becomes obvious that fragment detachment depends on the molecular weight even under conditions where the destruction of contact pairs is rate determining. Thus it is concluded that contact pairs are not decomposed in a concerted action but in a gradual process and that the lifetime of single contact pairs is of the order of 10 to 50 μs.

4. Studies Concerning Preferential Solvation

If a polymer is dissolved in a mixture of liquids the interaction of one liquid with the polymer might be much stronger than that of the others. In this case, the molecules of the strongly interacting compound are preferentially bound to the macromolecules. Frequently, preferential solvation is indicated by a change in intrinsic viscosity [Dondos (1971); Dondos and Benoit (1977)], and the difference $[\eta]_{ad}^{-1} - [\eta]_o^{-1}$ is taken as a measure for the extent of preferential solvation. According to Strazielle and Benoit (1961) the coefficient of preferential solvation is given by eq.(10):

$$\frac{\lambda'}{C} = \frac{1}{[\eta]_{ad}} - \frac{1}{[\eta]_o} \qquad (10)$$

$[\eta]_o$ and $[\eta]_{ad}$: intrinsic viscosity in the absence and presence of additive, respectively.

C: constant and $\lambda' = \frac{\gamma V}{M}$;

γ = excess number of preferentially adsorbed molecules
V = molar volume of preferentially adsorbed compound

With respect to fragment separation following main-chain rupture of linear polymers it was assumed that substitution of small solvating molecules by bulky ones might reduce the mobility of the fragments. Preferential solvation is illustrated in Fig.7, where two fragments of a macromolecule immediately after main-chain rupture are depicted.

In case (a) the fragments are totally solvated by small solvent molecules. Case (b) refers to the replacement of a few small molecules by bulky ones and case (c) to the total replacement of small molecules by bulky ones. When copolymers containing relatively small portions of photolytically degradable phenylvinyl ketone were irradiated with 347 nm light it turned out that preferential solvation was reflected by reduced rates of fragment separation if small solvating molecules were replaced by bulky ones. Typical results obtained with poly(N-vinylpyrrolidone-co-phenylvinyl ketone) containing 10 mol% PVK are presented in Fig.8.[Dobrowolski, Schnabel, unpublished]. Here, the copolymer was dissolved in chloroform and naphthol or phenol were added.

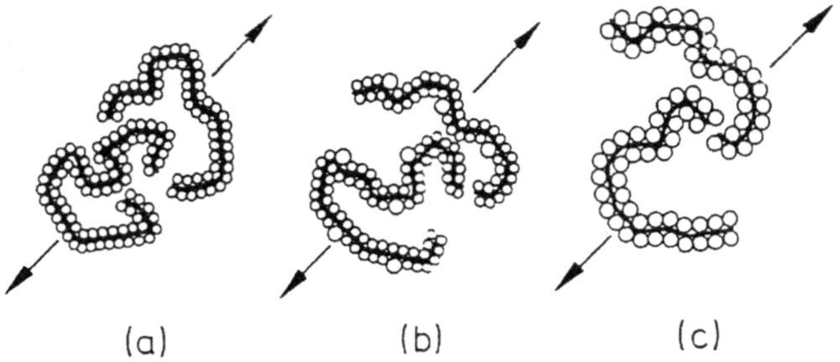

Fig.7. Schematic illustration of main-chain scission of a preferentially solvated macromolecule

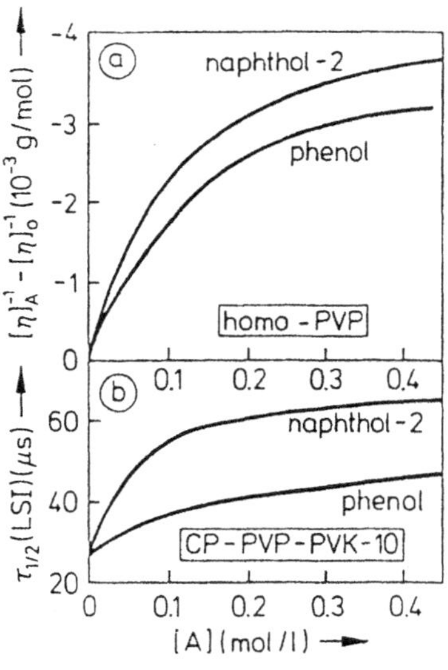

Fig.8. Preferential solvation of poly-N-vinylpyrrolidone by naphthol-2 and phenol. Change in the intrinsic viscosity (a) and variation in the rate of fragment separation (b) with additive concentration in chloroform solution at room temperature. (a): homopolymer ($M_{w,o}$= 1.1×10^6); (b): copolymer of vinylpyrrolidone and phenylvinyl ketone (PVK) containing 10 mol % PVK ($M_{n,o}$= 4.4×10^5). [Polymer]= 4×10^{-2}base mol/l.

In Fig.8(a) it is shown that the intrinsic viscosity increased with increasing additive concentration (note the negative sign of the difference $[\eta]_{ad}^{-1}-[\eta]_{o}^{-1}$), and from Fig.8(b) it is seen that the increase in $[\eta]$ is paralleled by an increase in $\tau_{1/2}$, i.e. by a decrease in the rate of fragment separation. Similar results were obtained upon irradiating poly(methylmethacrylate-co-phenylvinyl ketone) containing 7.8 mol % PVK in acetone solution upon the addition of naphthalene [Dobrowolski,Schnabel (1976)].

5. Studies Concerning the Dynamics of Polynucleotides

5.a. Detachment of DNA Double Strands

Polynucleotides such as native and denatured deoxyribonucleic acid (DNA) and synthetic poly(ribonucleic acids) (e.g.poly(riboadenylic acid), poly A, poly(ribocytidylic acid), poly C, etc.) undergo predominantly main-chain scission upon irradiation with high energy radiation in aqueous solution [see e.g. Bertinchamps (1978); von Sonntag, Schulte-Frohlinde (1978)]. Since native DNA is double-stranded fragmentation is expected for the case that both strands are ruptured on sites located opposite to each other. Actually, two single strand breaks (ssb's) can be the basis for a double strand break (dsb) if they are located on opposite strands within a distance h, as illustrated in Scheme 2. Here, h denotes the number of nucleotide units between single ssb's.

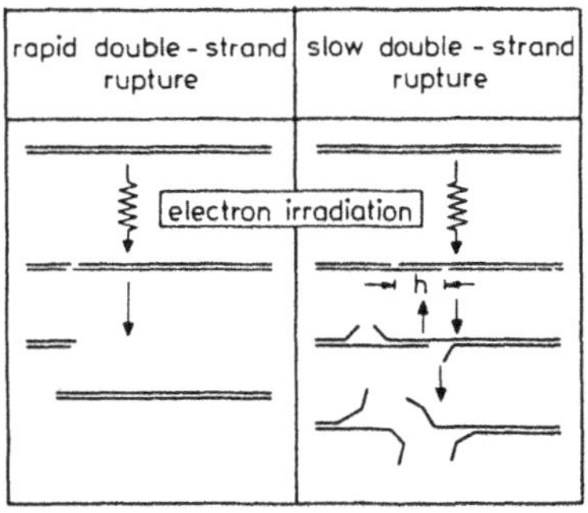

Scheme 2. Schematic illustration of the generation of double strand breaks (dsb's) in DNA

Fig. 9 shows kinetic traces demonstrating the decrease in LSI after irradiation of native calf thymus DNA in O_2-saturated aqueous solution. In this case ssb's are essentially generated by the attack of the sugar moieties of the DNA by OH radicals stemming from the radiolysis of water. The kinetic traces demonstrate that the LSI decreases in two different modes. The fast mode was attributed to the case where ssb's are located directly (or almost directly) opposite to each other and the slow mode to the case where the ssb's are separated by a distance h [Lindenau et al. (1976a) and (1976b)]. In other words, the rate of the fast mode was considered to reflect the rate of fragment diffusion whereas the rate of the slow mode was attributed to the rate of the detachment of segments located between ssb's. This assignment was greatly based on the temperature dependence of both the extent and the halflife of LSI decrease. In order to understand the results obtained, the model depicted in Scheme 1 had to be refined. It appeared appropriate to discriminate the distance h_f from the distance h_{crit}, where the subscripts "f" and "crit" refer to "fast" and "critical", respectively. Definitions of h_f and h_{crit} are given in Scheme 3.

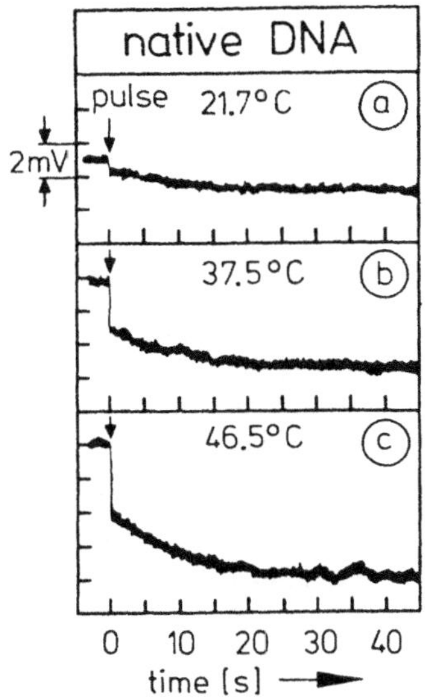

Fig.9
Main-chain degradation of native DNA induced by 16 MeV electron beam irradiation in oxygen-saturated 0.01 M NaCl solution. Kinetic traces demonstrating the fast and the slow mode of LSI decrease after the pulse at different temperatures. Absorbed dose per pulse: 500 Gy. Pulse length: 2 μs. [Lindenau et al (1976b)]

$h < h_f$: "fast" dsb's occur

$h < h_{crit}$: dsb's occur

$h_{crit} > h > h_f$: "slow" dsb's occur

Scheme 3. Schematic illustration concerning the definition of h_f and h_{crit}.

All dsb's contributing to the rapid process are due to ssb's separated by no more than h_f nucleotide units. On the other hand, there exists a critical number of units, h_{crit}, above which ssb's cannot become effective as dsb's.

The temperature effects observed during the investigations with native DNA are summarized below [Lindenau et al. (1976a; (1976b]:

(a) The total extent of LSI decrease was augmented upon increasing the temperature because of an increase in h_{crit}.
(b) $\tau_{1/2}$(slow) decreased with increasing temperature because of the higher rate of H-bond breakage at higher temperature.
(c) The fraction of the fast LSI decrease was augmented with increasing temperature. This is due to the fact that h_f is larger at higher temperatures.
(d) At T > 40°C a third mode of LSI decrease was detected, which presumably corresponds to the detachment of rather long fragments, which do not detach at all at lower temperatures.

Fig. 10 shows the increase of the fraction of the rapid mode of LSI decrease with increasing temperature. Notably, this curve resembles the "melting" curves of double stranded oligonucleotides indicating that the observed effect is related to the detachment of nucleotide segments of relatively short length. The inflection point of the curve in Fig. 10 refers to about 35°C. With respect to data concerning oligonucleotides [Pörschke, Eigen (1971)], this temperature appears to reflect the detachment of segments comprising about 7-15 nucleotide units.

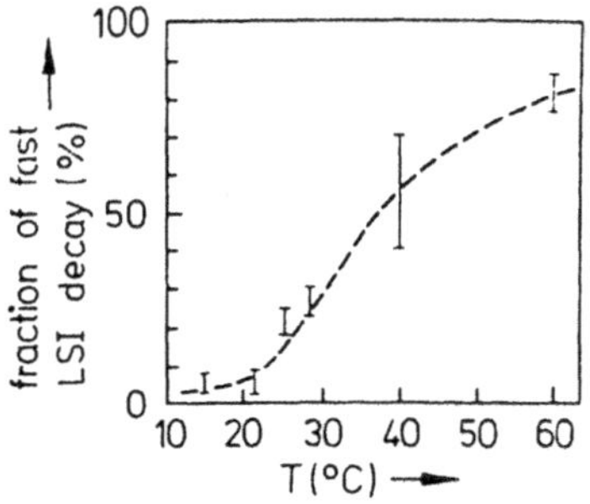

Fig.10. Main-chain degradation of native DNA. Fraction of fast mode of LSI decrease vs. temperature. [DNA]= 0.075 g/l in oxygen-saturated 0.01 M NaCl solution. Absorbed dose per pulse: 500 Gy. Pulse length : 2 μs. [Lindenau et al (1976b)]

5.b. Lifetime of Salt Bridges

Polyelectrolytes often form complexes in aqueous solutions upon the addition of multiply charged low molecular weight ions. These ions function as bridging agents between the polymer chains. Actually, salt bridges can be formed both intra- and intermolecularly.At sufficiently high salt concentrations the formation of intermolecular bridges results in the precipitation of the polymer. Notably, salt bridges are instable. However, there is a dynamic equilibrium between polymer-bound and free ions.
Investigations on the question concerning the average lifetime of salt bridges were carried out sometime ago in our laboratory with poly(riboadenylic acid) [Denk et al. (1983)].

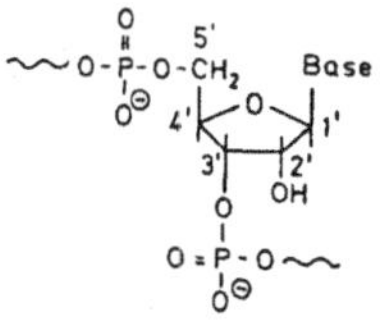

Base: 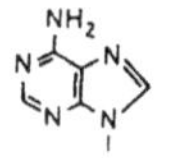

It is well known [Krakauer (1971)] that magnesium ions form complexes with poly A and other nucleic acids. In this case complex formation is due to salt bridges between phosphate groups:

$$O{=}P{-}O^{-}\ldots Mg^{2+}\ldots {}^{-}O{-}P{=}O$$

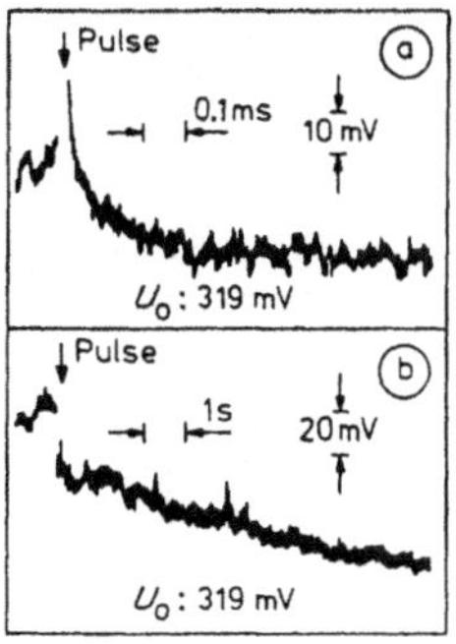

Fig.11
Main-chain scission of poly A in N_2O-saturated aqueous solution at pH 8 induced by the attack of OH radicals. Kinetic traces illustrating the decrease in LSI after the pulse at different time scales. Ionic strength: 0.1 mol/l. [poly A]= 0.2 g/l. $[Mg^{2+}]$= 0.012 mol/l. Absorbed dose per pulse 290 Gy. The solution was irradiated with 16 MeV electrons [Denk et al. (1983)].

When poly A was irradiated with a 100 ns pulse of 16 MeV electrons in aqueous solution containing Mg^{2+} ions (1.2×10^{-2}mol/l) at pH 8, the LSI decreased after the pulse due to main-chain scission. As can be seen from Fig.11 the decrease occurred in two modes. Two modes of LSI decrease were also observed in magnesium-free solutions. However, in the presence of Mg^{2+} ions the lifetime of the slow mode was found to be very much longer than in its absence [Washino et al. (1983)]. Regarding the rapid mode no difference in the lifetimes was found. Therefore, the rapid mode was assigned to main-chain breaks located in non-complexed parts of the macromolecules. The slow mode, on the other hand, was attributed to main-chain breaks located in complexed portions. Scheme 4 illustrates the two cases:

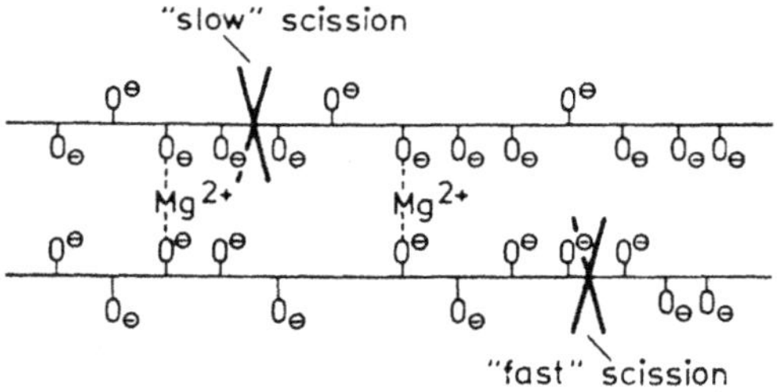

Scheme 4. Schematic illustration of the location of main-chain breaks leading to "slow" and "fast" separation of fragments. Poly A strands are depicted as straight lines showing only the phosphate oxygens with the negative charge [Denk et al. (1983)].

If main-chain scissions are located in complexed regions, separation of the fragments becomes possible only after the salt bridges in the neighborhood of broken bonds in the main-chain are ruptured. As has been pointed out before (vide ante) salt bridges have a limited lifetime. They are formed and destroyed in an equilibrium process. Provided the lifetime of the salt bridges exceeds the lifetime of the radicals decaying via main-chain rupture, decomposition of salt bridges is the rate-determining process with respect to the change in LSI observable in the presence of bivalent positive ions. Since this applies to the present case, it can be concluded that the slow mode of LSI decrease reflects the lifetime of the salt bridges, which has a magnitude of several seconds in the system poly A/Mg^{2+} under the conditions given in the legend to Fig.11.

6. Concluding Remarks

Concluding this article it should be pointed out again that main-chain scission of polymers as detectable by light scattering measurements can be utilized in conjunction with special techniques such as flash photolysis, pulse radiolysis or the stopped-flow method to obtain detailed insight in dynamic processes of polymers in solution.

R e f e r e n c e s

S.W.Beavan, W.Schnabel (1978), Macromolecules 12, 782

S.W.Beavan, G.Beck, W.Schnabel (1978), Eur.Polym.J. 14, 385

G.Beck, D.Lindenau,W.Schnabel (1974), Eur.Polym.J. 10, 1069

G.Beck, D.Lindenau, W.Schnabel (1975), Eur.Polym.J. 11, 761

G.Beck, D.Lindenau, W.Schnabel (1977), Macromolecules 10, 135

G.Beck, G.Dobrowolski, J.Kiwi, W.Schnabel (1975), Macromolecules 8,9

G.Beck, G.Dobrowolski, J.Kiwi, W.Schnabel (1975), "Lasers in Physical Chemistry and Biophysics", Elsevier Scientific Publ., Amsterdam, p.283

A.J.Bertinchamps (ed.) (1978), "Effects of Ionizing Radiation on DNA", Springer-Verlag, Berlin

K.-J.Deeg, L.Katsikas, W.Schnabel (1984), Z.Naturforsch. 39c, 1075

K.J.Deeg, L.Katsikas, W.Schnabel (1985), Helv.Chi,Acta 68, 2367

K.J.Deeg, L.Katsikas, W.Schnabel (1987), Int.J.Radiat.Biol. 51, 527

O.Denk, K.Washino, W.Schnabel (1983), Makromol.Chem. 184, 165

G.Dobrowolski, W.Schnabel (1979), Z.Naturforsch. 31a, 109

G.Dobrowolski,J.Kiwi, W.Schnabel (1976), Eur.Polym.J. 12, 657

A.Dondos (1971), C.R.Acad.Sc. Paris, Ser.C,272

A.Dondos, H.Benoit (1977), J.Polym.Sci.,Polym.Phys.Ed. 15, 137

U.Gröllmann, W.Schnabel (1980), Makromol.Chem. 181, 1215

K.Horie, W.Schnabel (1982), Polym.Photochem. 2,419

K.Horie, W.Schnabel (1984), Polym.Degrad.Stabil. 8, 145

H.Kihara, W.Schnabel (1980), Z.Naturforsch. 35a,217

J.Kiwi, W.Schnabel (1975), Macromolecules 8, 430

H. Krakauer (1971), Biopolymers 10, 2459

D.Lindenau, S.W.Beavan, G.Beck, W.Schnabel (1977), Eur.Polym. J. 13, 819

D.Lindenau, U.Hagen, W.Schnabel (1976a), Z.Naturforsch. 31c, 484

D.Lindenau, U.Hagen, W.Schnabel (1976b), Rad.Environm.Biophys. 13, 287

D.Lindenau, W.Schnabel (1978), J.Polym.Sci., Polym.Lett.Ed. 16, 655

D.Pörschke, M.Eigen (1971), J.Mol.Biol. 62, 361

J.Rosiak, W.Schnabel (1984), Eur.Polym.J. 20, 1159

J.Rosiak, W.Schnabel (1986), J.Radioanal.Nucl.Chem. 101, 433

W.Schnabel (1979),"Application of the light scattering detection method to problems of polymer degradation" in N.Grassie (ed.) "Developments in Polymer Degradation", Applied Science Publ., London, Vol.2

W.Schnabel (1986), Radiat. Phys.Chem. 28, 303

W.Schnabel (1988), Polym.News 13, 199

H.Seki, W.Schnabel (1982), Z.Naturforsch. 37c, 63

C.v.Sonntag, D.Schulte-Frohlinde (1978), in "Effects of Ionizing Radiation on DNA", A.J.Bertinchamps (ed.), Springer-Verlag, Berlin, p. 204

C.Strazielle, H.Benoit (1961), J.Chim.Phys. 58, 675

S.Tagawa, W.Schnabel (1979 a), Macromolecules 12, 663

S.Tagawa, W.Schnabel (1979 b), Makromol.Chem. 180, 2371

S.Tagawa, W.Schnabel (1983), Makromol.Chem., Rapid. Commun. 3, 427

K.Washino, W.Schnabel (1980), Z.Naturforsch. 35a, 637

K.Washino, W.Schnabel (1982a), Makromol.Chem. 183, 697

K.Washino, W.Schnabel (1982b), Int.J.Radiat.Biol. 41, 271

K.Washino, W.Schnabel (1982c), Makromol.Chem.,Rapid Commun. 3, 427

K.Washino, O.Denk, W.Schnabel (1983), Z.Naturforsch. 38c, 100

K.Washino, L.Katsikas, W.Schnabel (1984), Int.J.Radiat.Biol. 46, 747

CARBON 13 NMR OF GAMMA-IRRADIATED POLYETHYLENES

James C. Randall
Machelen Chemical Technology Center
Exxon Chemical International
Nieuwe Nijverheidslaan 2
B-1920 Machelen
Belgium

ABSTRACT. Carbon 13 NMR has been used to investigate the structural changes occurring in polyethylene as induced by either ^{60}Co or electron beam irradiations. In this brief review, NMR results are compared for solution and solid state NMR studies. There is a problem with the available sensitivity from both types of carbon 13 NMR techniques, although considerable information has become available from both the solution and solid state NMR studies. Chain scission, the formation of cis and trans double bonds in the amorphous phase, the formation of trans double bonds in the crystalline phase and the formation of "Y" types of long chain branches have been observed from these NMR studies. As background information for the quantitative carbon 13 NMR study of the radiation of polyethylene, a brief examination of polyethylene molecular weight measurements using carbon 13 NMR on NBS 1475 and a comparison of the structures of a medium density ethylene-1-hexene copolymer before and after severe melt processing are included.

1. Introduction

Carbon 13 nuclear magnetic resonance (NMR) has revealed the molecular structure of polymers in a way unparalleled by any other technique. A sensitivity to differences in chirality among polymer repeat unit sequences to as many as nine contiguous units has led to detailed information regarding the configurational distributions in polymers(1). Triad distributions, which lead to average sequence lengths(1), run numbers(2) and sequence numbers(3), can be routinely measured for a broad array of copolymers. With the sensitivity provided by the newer, more advanced superconducting magnet systems and accompanying improved software, one carbon in ten thousand can be detected for those polymers yielding spectra with narrow line widths(4). Thus carbon 13 NMR has become a reasonable technique to measure the number average molecular weight of polyethylene(4)(5). With these advantages in mind, it is the purpose of this article to review the information carbon 13 has revealed in studies of structural changes induced by gamma irradiation of polyethylene.

O. Güven (ed.), Crosslinking and Scission in Polymers, 57–82.

The often published phenomenological descriptions of nuclear magnetic resonance will not be repeated here, although it may be instructive to review briefly a few key attributes that have made carbon 13 NMR so amenable to the determination of polymer structure. One advantage carbon 13 NMR has over corresponding NMR studies involving other nuclei is that carbon 13 NMR has a natural abundance of only 1%. With all proton interactions broad band decoupled, carbon 13 NMR spectra yield predominantly chemical shift information with each line representing some type of structural moiety detected by the carbon 13 NMR technique. This presents a great advantage in structural studies when radiation-induced structural changes are being identified. A simple comparison of NMR spectra before and after irradiation can reveal both the extent and type of changes that have occurred as a result of the radiation treatment. The fact that the carbon 13 nucleus has a nuclear spin of 1/2 presents another advantage in that quadrapole effects are absent thus making narrow line widths possible. The spectra can also be treated theoretically in a similar way to proton NMR. Carbon 13 NMR resonances cover a range of 0-500 ppm with the majority from hydrocarbon polymers falling in a range of 0-200 ppm. This is still twenty times greater than the range of corresponding proton NMR chemical shifts for the same structural moieties.

Advances in instrument technology is another factor contributing to the success of carbon 13 NMR in structural analyses of polymers. Operation in the Fourier transform mode has made high sensitivity possible and allows a trade-off between sensitivity and resolution whenever desirable(6). Broad band decoupling, also results in the nuclear Overhauser effect (NOE), which leads to an enhancement of the carbon 13 resonance intensities. The magnitude of the NOE is related to the specific mechanism through which the carbon 13 nuclear spins relax after excitement by a radiofrequency field. It is fortunate that most polymers relax almost exclusively through a dipole-dipole mechanism as a result of restricted mobility. Consequently, a maximum intensity enhancement of 3.0 is observed for the NOE(7) and meaningful quantitative analyses generally go unimpeded.

2. Carbon 13 NMR of Polyethylene

Polyethylene has excellent NMR features when looking for structural changes as a consequence of a given process because of the simplicity of its carbon 13 NMR spectra. Most of the carbons in high density polyethylene are in exactly the same environment with structural uniqueness occurring only for those carbons near the ends of the chain or associated with branches. Consequently, the majority of the carbons in polyethylene resonate at approximately 30 ppm from an internal tetramethylsilane standard. Figure 1 contains a carbon 13 NMR spectrum of NBS 1475 as observed at 50 MHz in 1,2,4-trichlorobenzene and 125°C. The major resonance occurs at 29.98 ppm and is labeled $\delta^+\delta^+$.

The Greek alphabet is used to identify the structural location of a methylene carbon by indicating alphabetically the distance in carbon atoms to the nearest branch in both directions. For example, a methylene carbon sandwiched between two branched carbons is labeled $\alpha\alpha$. The methylene carbons giving rise to the 29.98 ppm resonance are labeled $\delta^+\delta^+$ to define the fact that the nearest branch is four or more carbons removed in either direction. This also indicates that structural uniqueness only occurs for carbons that are within four carbons from a branch, end group or some structural irregularity. Carbons in a branch are labelled nB_i where "n" designates the position in the branch and "i" gives the branch length. The methyl carbon at the end of the branch is labelled "1" and the numbering starts towards the branched carbon in the polymer backbone. The carbons in the chain ends are labelled 1s, 2s, 3s, etc. starting with the methyl end group. (The symbol, "s", stands for "saturated".) If one of the end groups is a terminal double bond, the allylic carbon is designated as "a". Table I contains chemical shifts and intensity information for NBS 1475 polyethylene.

TABLE I

Carbon 13 NMR Resonance Areas and Peak Heights for NBS 1475 Polyethylene from a 50 MHz Spectrum taken at 15 Wt. % in 1,2,4-Trichlorobenzene at 125°C (See Figure 1.)

		Resonance Intensities	
Carbon	δ, TMS	By Spectral Integration	By Peak Height
1s	14.15 ppm	0.914	40.5 mm
2s	22.86	0.876	50.7
3s	32.18	1.043	56.9
$\delta^+\delta^+$	29.98	902.402	30,000.0
a	33.91	0.484	19.4

The terminal vinyl carbon resonances occur at 114.1 ppm (1"v") and 139.2 ppm (2"v") respectively from TMS.

There are several key features to be noted in Figure 1 and from the data in Table I. There is a difference by nearly three orders of magnitude between the peak heights for $\delta^+\delta^+$ and the end group resonances. For example, $\delta^+\delta^+$ is 30 *meters* high while the end group resonances range from 40-60 *millimeters* in peak heights. Consequently, double precision arithmetic must be used during data accumulation to avoid truncation. Secondly, the relative intensities of the NMR resonances 1s, 2s and 3s are expected to be the same. The relative areas display the expected similarities for 1s and 2s, but the peak heights differ by at least 20%. An enhanced peak height and area for 3s relative to 1s and 2s results from a contribution from the baseline tail from the major $\delta^+\delta^+$ resonance. Each of the resonances, including $\delta^+\delta^+$, have a peak width at one-half height of approximately one hertz. Slight differences here may also cause peak height variations. Narrow linewidths in this range are a requirement if a sensitivity of one carbon in 10,000 carbon atoms is to be achieved.

Sufficient information is available for NBS 1475 to determine the number average molecular weight by ^{13}C NMR by simply taking a ratio of the total carbon resonance intensity to the intensity observed for a single carbon, that is,

$$Mn = [\text{Total Carbon Intensity/A Single Carbon Intensity}] \times 14. \quad (1)$$

The single carbon intensity can be determined from the end group resonance intensities as follows:

$$\text{A single Carbon Intensity} = (1/2)[\bar{s} + a] \quad (2)$$

and the number average molecular weight is given by,

$$Mn = \{(1s + 2s + 3s + \delta^+\delta^+ + 3a) \times 14\}/(1/2[\bar{s} + a]) \quad (3)$$

If an average ($\bar{s}$) of 1s, 2s, and 3s cannot be used because of baseline or relaxation problems, it is best to pick the most reliable observed value and use it consistently throughout where an "s" or $\bar{s}$ is needed.

The National Bureau of Standards reports the number average molecular weight for NBS 1475 as 18,300(8). Both the peak height data and peak area data from Table I can be used independently to determine the number average molecular weight of NBS 1475 by carbon 13 NMR. This will establish just how serious the baseline and linewidth problems may be. Substituting the relative area data into Equation 3 gives an Mn of 17,800 while the peak height data gives a value of 12,300. An improved result for the Mn from area measurements is obtained if an $\bar{s}$ from 1s and 2s is used as opposed to an $\bar{s}$ from 1s, 2s and 3s. It becomes quite clear that relative areas are preferred over peak heights for quantitative measurements in polyethylene. Small changes in linewidths can create substantial differences in peak heights even though the relative areas are the same. There are still studies where extremely weak and noisy resonances do not permit reliable area measurements. A peak height may have to be used, but it should be realized there may be a substantial error in the measurement and it should only be used to establish trends or differences among samples.

3. Severe Melt Processing of Polyethylene

The good agreement between number average molecular weights determined from carbon 13 NMR peak area measurements and that published for NBS 1475 indicate that carbon 13 NMR has the capability to yield valuable quantitative information in studies involving irradiation and thermal degradation of polyethylene. It may be instructive at this point to examine an ethylene-1-hexene copolymer before and after a deliberate severe melt processing in a Brabender extruder(4). Carbon 13 NMR affords the opportunity to establish the structural changes that occur and to note just how much each carbon resonance intensity is affected by severe melt processing. Carbon 13 NMR spectra at 50 MHz of a medium density ethylene-1-hexene copolymer before and after severe melt processing are shown in Figure 2. The spectra are presented under precisely the same solution and NMR instrumental conditions. Both samples are at 15 weight percent and the spectra were taken at 125°C. The appearance of resonances associated with long chain branching is unmistakably evident. Substantial intensity increases have occurred for the methine resonance and for both $\alpha\delta^+$ and $\beta\delta^+$, while the side chain carbon resonances for $2B_4$ and $4B_4$ are hardly affected. The allylic carbon resonance, "a", has decreased in intensity showing a loss of terminal vinyl groups. Resonances associated with hydroperoxide and carbonyl groups appear also. The methine carbons are expected to be more responsive to oxidation than the main chain methylene carbons, but, in this case, there appears to be little, if any, effect from autoxidation. The lack of change in the $2B_4$ and $4B_4$ branch carbon chemical shifts and intensities can be used to indicate that very little oxidation occurred at the methine carbons. The most likely explanation is that the relative population of butyl branches is so much less than that of the main chain methylene carbons that severe melt processing had little effect on the observed branch carbon populations. Results from NMR quantitative measurements before and after severe melt processing are given in Table II.

TABLE II

The Average Structural Compositions of a Medium Density Ethylene-1-Hexene Copolymer Before and After Severe Melt Processing. The Data is Determined from Integrated Resonance Intensities of the Spectra Shown in Figure 2.

Structural Unit	Number of Structural Units per 10,000 Carbon Atoms	
	Before	After Severe Processing
Saturated End Groups	9.8	10.1
Terminal Vinyl End Groups	9.6	4.1
Butyl Branches	6.8	6.8
Carbonyl Groups	Negligible	4.2
Hydroperoxide Groups	Negligible	2.4
Long-Chain Branching*	Negligible	3.5

* Determined from the $\alpha\delta^+$ resonance area after correction for butyl branches.

The formation of the "Y" type of long chain branch is clearly evident by the enhancements in intensities for the methine, $\alpha\delta^+$and $\beta\delta^+$ resonances. The mechanism for its formation is not so clear. Chain scission followed by recombination with nearby secondary main chain radicals may be responsible for some long chain branch formation. It may also be possible that some of the terminal vinyl groups are combining with the main chain secondary radicals(9). It is interesting to note that the loss in terminal vinyl groups during processing accounts for only about 60% of the observed amount of long chain branching formed. The peroxide and carbonyl groups could be established as main chain groups by their characteristic $\alpha\delta^+$ and $\beta\delta^+$ chemical shifts, that is,

$$\overset{24.3}{} \quad \overset{42.8}{} \qquad\qquad\qquad\qquad\qquad\qquad\qquad \overset{33.1}{} \quad \overset{26.8}{}$$

$$-CH_2-CH_2-CH_2-\underset{\underset{O}{\|}}{C}-CH_2-CH_2-CH_2- \qquad -CH_2-CH_2-CH_2-\underset{\underset{\underset{H}{|}}{\underset{O}{|}}}{\underset{O}{|}}{CH}-CH_2-CH_2-CH_2-$$

as opposed to terminal carbonyl or hydroperoxide groups possibly resulting from olefin oxidation. An appreciable amount of chain scission would be expected to produce saturated and groups after chain transfer. The perceptible, but small observed increases in 2s and 3s are within experimental error of the NMR measurement.

The following observations could be established as the most important results of severe melt processing of a medium density polyethylene:

(a) long chain branches form upon severe melt processing,
(b) the number of terminal vinyl groups reduce substantially after severe melt processing,
(c) carbonyl groups and hydroperoxide groups form during or after severe melt processing,
(d) there are no apparent changes in original short chain branch concentration with severe melt processing of a medium density polyethylene and
(e) there is also no significant change in the saturated end group concentration after severe melt processing.

This brief study on the structural changes from the free radical chemistry induced by severe melt processing serves as a good introduction to a subsequent study of the similar radiation-induced structural changes in polyethylene. As stated previously, polyethylene is an ideal system for carbon 13 NMR studies of the chemical changes caused by either processing, ageing or irradiation because most of the carbons resonate uniquely near 30 ppm from an internal TMS standard.

4. Irradiation of Polyethylene - Solution NMR

Both NBS 1475 polyethylene and Marlex 6003 have been examined carefully for structural changes occurring after ^{60}CO gamma irradiation (10)(11). As discussed in the preceding sections, NBS 1475 has been previously well characterized. Marlex 6003 was selected because there are two types of end groups, a saturated end group and a terminal vinyl end group in nearly equal proportions. As shown in Table I, only about one third of the end groups in NBS 1475 are terminal vinyl. The Marlex 6003 was obtained in a nascent granular form and contained no additive package. In contrast, NBS 1475 contains a stabilizer package. This difference should be kept in mind when comparing the responses to radiation by these two polyethylenes.

Before examining carbon 13 NMR polyethylene spectra taken before and after ^{60}Co irradiation, some additional chemical shift information is needed. The "Y" branch has been well characterized, as have the chemical shifts associated with the various types of branches and end groups (4). What is needed is chemical shift information associated with the so-called "H" link, the structural moiety expected from a coupling of two main chain secondary radicals, that is,

$$\begin{array}{l} \quad\;\;\gamma \quad\;\; \beta \quad\;\; \alpha \\ -CH_2-CH_2-CH_2-CH-CH_2-CH_2-CH_2- \\ \qquad\qquad\qquad\quad\;\; | \\ -CH_2-CH_2-CH_2-CH-CH_2-CH_2-CH_2- \end{array}$$

So far, the only unequivocal identification of an "H"-link has been after gamma irradiation of n-alkanes (12,13). Bennett, et al., (12) established the "H"-link chemical shifts from the model compound, 1,1,2,2-tetra (tridecyl)ethane and observed "H"-link formation following an irradiation of n-hexadecane and n-eicosane. The "H"-link methine and methylene chemical shifts reported by Bennett, et al., are given below:

$$\begin{aligned} \alpha &= 30.73 \text{ ppm} \\ \beta &= 28.22, \\ \gamma &= 30.19, \\ \text{Methine} &= 39.49. \end{aligned}$$

Bovey, Schilling and Cheng (13) observed similar "H"-link chemical shifts after irradiation of a $C_{44}H_{90}$ linear alkane in the molten state. "Y" branch resonances and resonances from trans double bonds were observed as well.

The observed "H"-link chemical shifts reported in the literature may vary somewhat depending upon the choice of solvent and sample temperature during the NMR experiment. These model compound studies are therefore useful only for establishing the general vicinity of the "H"-link resonances in the NMR spectrum of polyethylene after irradiation. More precise information requires a repetition of the model compound work and chemical shift observations under the NMR conditions normally used for polyethylene.

Bovey, et al (13) reported that irradiation of a $C_{44}H_{90}$ linear alkane in the crystalline state produced no obvious effects in the ^{13}C NMR spectrum, although GPC analyses suggested that about

1% of the molecules had doubled in molecular weight. It was concluded that "end-linking" occurred through terminal methyl groups in either adjacent rows or adjacent layers in the crystal structure.

In an analogous radiation and subsequent ^{13}C NMR study of polyethylene, Bovey and coworkers were unable to observe NMR resonances from either the "H"-link or the "Y" branch. Generally, little success has been encountered in ^{13}C NMR characterization studies of crosslinked polyethylenes. The spectra are noisy even at high polyethylene concentrations, suggesting that only the highly mobile segments give rise to observable NMR resonances. To avoid such difficulties, it may be best to examine polyethylenes that are irradiated just up to the gel dose (10)(11). It is also necessary to establish the reference "H"-link chemical shift information under NMR instrumental conditions generally utilized to obtain ^{13}C NMR spectra of polyethylenes.

4.1 CARBON 13 NMR OF n-HEXATRIACONTANE

n-Hexatriacontane ($C_{36}H_{74}$) was irradiated at 100 Mrad in both the melt and at room temperature to establish the "H"-link chemical shifts in 1,2,4-trichlorobenzene and at 125°C (10). Results are given in Table III. The only completely obvious change after irradiation in the solid state was the appearance of resonances from trans double bonds with an allylic carbon resonance appearing at approximately 32.5 and a trans-olefinic carbon resonance near 131 ppm. Irradiation in the melt produced resonances for "Y" branches, cis and trans double bonds and new resonances that correspond reasonably well to those reported for the "H"-link. The data in Table III is expressed in number of structural units per 10,000 carbons.

TABLE III

n-Hexatriacontane Structural Unit Concentrations[1] Following 100 Mrad ^{60}Co Irradiation under Vacuum

	100 Mrad ^{60}Co Irradiation Number of Units per 10,000 Carbons	
Structural Unit	25°C	82°C (Melt)
Saturated End Groups	562	555
Trans Double Bonds	8.6	4.9
Cis Double Bonds	None Detected	1.5
"Y" Branches	None Detected	5.4
"H" Links	Trace	3.3

[1]The saturated end group concentration of n-Hexatriacontane is 556 per 10,000 carbon atoms.

The NMR data in Table III for n-hexatriacontane was obtained from a saturated solution in perdeuterobenzene to permit an observation of the olefinic carbon resonances. It was still possible to obtain data at 125°C, the temperature of polyethylene NMR measurements. A resonance at 41.01 ppm was confirmed to originate from a methine carbon by off-resonance decoupling. The identification of the α, β and γ methylene resonances was made with less certainty because of the proximity to the major $\delta^+\delta^+$ resonance. Resonances at 30.5, 28.6 and 30.6 ppm were assigned to the α, β and γ methylene carbons of the H-link, respectively. Similar chemical shipts were observed in 1,2,4-trichlorobenzene. The difference in the methine shift assigned to the H-link differs by approximately 2 ppm from that reported by Bennett, et al. (12). By lowering the temperature to 50°C and preparing the sample in a mixture of perdeuterochloroform and 1,2,4-trichlorobenzene, the methine resonance assigned to the "H"-link shifted from 41 to 39.95 ppm, which is closer to the value reported by Bennett, et al., for their model compound dissolved in perdeuterochloroform and under ambient probe conditions. Thus under the NMR experimental conditions required for polyethylene, the "H" link methine resonance is expected to reside closer to 41 ppm. Sufficient confirmation has been obtained to identify the region in the ^{13}C NMR spectrum of polyethylenes where the H-link resonance is expected to resonate following ^{60}Co gamma irradiations.

4.2 CARBON 13 NMR SPECTRA OF NBS 1475 BEFORE AND AFTER ^{60}Co GAMMA IRRADIATION.

NBS 1475 polyethylene was irradiated under the following conditions : 2.0, 4.0 and 8.0 Mrad at 25°C under vacuum, 4.0 Mrad in air and 1.0 Mrad under vacuum at 225°C after heating under vacuum for 24 hrs at 225°C, which led to considerable thermal degradation. Carbon 13 NMR spectra were obtained under exactly the same sample and NMR experimental conditions before and after the irradiations so that precise comparisons could be made. The polymers were dissolved in 1,2,4-trichlorobenzene at 15 weight % and the spectra obtained at 125°C. The 2.0 and 4.0 Mrad irradiations were made below the gel dose while the 8.0 Mrad treatment was performed to provide a sample irradiated above the gel dose. (It was the only sample to show a measurable gel content.)

The ^{13}C NMR spectrum of the sample irradiated to 8.0 Mrad and results have been reported previously(10). There are two significant observations. First, the only clearly obvious change from a spectral viewpoint was the appearance of resonances associated with the presence of trans internal double bonds. Secondly, the noise level of the spectrum was too high for the 15% concentration and number of FID's accumulated (see Table V). It is apparent that this NMR sample behaved as if it were more dilute. This result would occur only if the soluble portions of the sample were giving narrow resonances while the solvent swollen, gelled portions were not giving easily observable NMR resonances under the acquisition conditions of the NMR experiment. It is well known that NMR resonances broaden as the weight percent polymer in solution increases(4). It is possible that crosslinked polyethylene chains give signals so broad that they cannot be easily observed under normal NMR conditions for obtaining high resolution NMR spectra.

The samples irradiated below the gel dose provide an opportunity to detect structural changes occurring early on in polyethylene before gelled polymer is formed. Results are given in Table IV. Figure 3 contains spectra of NBS 1475 acquired before and after a 2.0 Mrad irradiation in the solid state under similar NMR sample and experimental conditions. The allylic carbon resonance for terminal vinyl end groups decreases in intensity while the appearance of resonances for the formation of "Y" branches is clearly evident. Resonances associated with internal main chain carbonyl groups can be easily identified also. There is a small, but perceptible increase in the saturated end group concentration. The cis and trans double bond contents were measured from the respective allylic carbon resonances, designated as a_t and a_c in Figure 5.

There is some scatter in the data in Table IV as should be expected with measurements in the range of 1-3 per 10,000 carbon atoms. Peak heights were necessarily used in a number of instances. The absolute values for the quantitative results derived from peak height results could be in error by as much as 50 to 100%, but the changes relative to given experiments should definitely indicate the directions of the results. For this reason, G values were not determined. The peak height measurements were made from resonances with similar linewidths as shown in Table V.

It is clear from the data in Table IV that the concentration of saturated end groups increases with irradiation, the concentration of vinyl end groups diminishes with irradiation and there is an increase in both the internal double bond content and "Y" type of long chain branch with increasing radiation doses up to the gel dose.

The molecular weight data, as determined through gel permeation chromatography, clearly indicate that linking reactions have occurred during the solid state ^{60}Co irradiations at room temperature. The 2.0 and 4.0 Mrad irradiations led to substantial increases in the weight average molecular weight and a broadening of the molecular weight distribution of NBS 1475. The number average molecular weights increased by approximately 20%. The observed weight average molecular weights in the range of 120,000 to 130,000 indicate that some molecules have doubled in size, suggesting that some linking reactions have taken place without any prior chain scission. Whether "Y" branch formation is responsible or end-linking has occurred, as suggested by Bovey, et al., (13) remains undecided. The increase in the saturated end group concentration is clear evidence that some chain scission reactions followed by chain transfer have also occurred. Inspite of this, the linking reactions were sufficient to affect dramatically the observed molecular weight averages and distribution.

The question remaining is whether or not "H" links form at these radiation levels or whether the NMR technique fails to detect the "H" link for the reasons cited earlier. To gain further insights, a thermal degradation/1.0 Mrad irradiation experiment was performed.

TABLE IV

Structural Concentrations Before and After Irradiation of NBS 1475 Under Vacuum

	Number of Units per 10,000 Carbons				
		2.0	4.0	8.0	1.0
	Before	Mrad.	Mrad.	Mrad.[1]	Mrad.[2]
Structural Unit		Irrad.	25°C	25°C	225°C
Sat'd. End Groups[3]	10.4	13.0	12.9	15.1	94.3
Vinyl End Groups[3]	5.3	2.0	2.8	---	16.1
"Y" Branches	0.7	0.9	1.0	1.3	43.5[3]
Cis Double Bonds	1.4	3.6	1.6	2.9	1.9
Trans Double Bonds	1.4	2.7	1.5	5.1	2.6
Ethyl Branches	2.5	3.2	4.0	4.3	4.1
Butyl Branches	N.D.[4]	N.D.	N.D.	N.D.	2.5
Hydroperoxide Groups	1.8	2.7	1.2	4.6	2.5
Carbonyl Groups	N.D.	2.2	1.3	N.D.	1.3
$M_w \times 10^{-3}$	52.8[5]	116[5]	128[5]	---	35.8[5]
$M_n \times 10^{-3}$	18.1	21.8	22.3		5.5
M_w/M_n	2.9	5.3	5.7		6.5

[1]Sample partially gelled

[2]Sample heated in vacuum at 225°C for 24 hrs before 1.0 Mrad irradiation

[3]Concentrations determined from peak areas; no designation indicates a peak height was used.

[4]N.D. = None Detected

[5]Measured by GPC

TABLE V

Total Number of Free Induction Decay (FID) Transients Collected and ^{13}C NMR Linewidths at 1/2 Peak Heights for the $\delta^+\delta^+$ Resonances for NBS 1475 Before and After Various Radiation Doses

Sample	Radiation Dose	$\delta^+\delta^+$ Line-Width at 1/2 Ht.	FID's Collected
NBS 1475	None	1.2 Hz	9,699
NBS 1475	2.0 Mrad at 25°C	1.2	4,482
NBS 1475	4.0 Mrad at 25°C	1.3	14,500
NBS 1475	8.0 Mrad at 25°C[1]	1.1	5,104
NBS 1475	1.0 Mrad + Thermal Degradation at 225°C	1.4	7,271

The NBS 1475 sample was heated in a vacuum at 225°C for 24 hrs prior to irradiation. A subsequent ^{13}C NMR scan revealed that considerable chain scissions had occurred by the level of saturated end groups. "Y" branches and terminal vinyl end groups had also formed in copious amounts. A subsequent irradiation to 1.0 Mrad resulted in a decrease in the terminal vinyl content, an increase in the "Y" branch content but in no obvious "H" links. See Table IV for the final quantitative results. No detectable change occurred in the saturated end group content. A comparison between the thermally degraded/1.0 Mrad irradiation of NBS 1475 and an ethylene-1-octene copolymer is shown in Figure 4. The similarities make it unmistakably clear that the irradiated NBS 1475 sample has very high levels of long chain branches. (A hexyl branch and a long chain branch give rise to similar NMR spectral patterns(4)). The sample has no gel content and there is only a barely perceptible resonance in the region for the methine carbon for the "H" link. (See the resonance near 40 ppm in Figure 4.) It is clear that under these conditions that the "H" link did not form to any appreciable extent. The formation of butyl branches is anticipated from the free radical chemistry associated with thermal degradation and is not a product of irradiation. These results become more apparent with the data presented later in Table VIII from the corresponding experiment from Marlex 6003. (Complete data and spectra for only one of the latter experiments are included as corresponding results were obtained.)

4.3 CARBON 13 NMR SPECTRA OF MARLEX 6003 BEFORE AND AFTER ^{60}Co GAMMA IRRADIATION

Marlex 6003 was irradiated under the following conditions : 2.0 Mrad in vacuum, 4.0 Mrad in air and 3.0 Mrad in vacuum after 24 hours of thermal degradation under vacuum at 225°C. Once again, ^{13}C NMR spectra were obtained under precisely the same experimental conditions so comparisons could be made before and after irradiation. The 2.0 Mrad irradiation was below the gel dose while a 4.0 Mrad irradiation in vacuum leads to a low production of gel.

Figure 5 contains ^{13}C NMR spectra of Marlex 6003 before irradiation and after a 2.0 Mrad dose. In the reference spectrum taken prior to irradiation, it is clear that the terminal vinyl end groups and the saturated end groups are present in similar amounts as shown by the corresponding intensities for "a" and the "2s" resonances in Figure 5. Irradiation in the solid state at 25°C to only an extent of 2.0 Mrad leads to a clear loss in the "a" intensity and the onset of resonances associated with "Y" branches. Note the increase in methine, $\alpha\delta^+$ and $\beta\delta^+$ resonance intensities in Figure 5. The quantitative data in Table VI give a clearer picture of the structural changes occurring with irradiation.

Once again, the use of peak heights in the quantitative measurements could lead to errors in the absolute concentrations by as much as 50 to 100%. The directional changes in the relative concentrations, however, should be quite reliable. Once again, the peak widths at 1/2 the peak height were determined to indicate the reliability of peak height measurements. The data is given in Table VII.

TABLE VII

Line-Widths at 1/2 the Peak Height of the $\delta^+\delta^+$ Resonances for Marlex 6003 Following Various Radiation Doses

Sample	Radiation Dose	Line-Width at 1/2 Ht.	FID's Collected
Marlex 6003	None	1.3	20,683
Marlex 6003	2.0 Mrad at 25°C	1.6	9,617
Marlex 6003	4.0 Mrad at 25°C in Air	1.0	5,250
Marlex 6003	24 h. Thermal Degradation at 225°C	1.3	4,314
Marlex 6003	24 h. Thermal Degradation at 225°C + 3.0 Mrad, 225°C	1.2	9,302

Molecular weight measurements also lend understanding to the radiation process. Irradiation to 2.0 Mrad in vacuum at room temperature leads to a substantial loss in terminal vinyl end groups while the number of saturated end groups and long chain branches increases. Correspondingly, the M_w increases from 140,000 to 164,000. The increase in saturated end groups undoubtedly is a result

TABLE VI

Structural Changes Before and After Irradiation of Marlex 6003 in Vacuum

	Number of Units per 10,000 Carbons		
	Before	2.0	4.0
	Irrad.	Mrad.	Mrad. in Air
Structural Unit	25°C	25°C	25°C
Sat'd. End Groups[1]	7.3	8.6	10.1
Vinyl End Groups[1]	7.8	3.9	4.1
"Y" Branches	1.2	2.2	N.D.[2]
Trans Double Bonds	1.7	3.1	3.1
Cis Double Bonds	1.7	2.5	3.1
Hydroperoxide Groups	1.8	2.8	4.6
Carbonyl Groups	0.7	N.D.	1.7
$M_w \times 10^{-3}$	140[3]	164[4]	58[4]
$M_n \times 10^{-3}$	20	---	---
M_w/M_n	7	>7	>4

[1]Concentration determined from relative areas. No designation indicates a peak height was used.

[2]N.D. = None Detected

[3]Measured by GPC

[4]Measured by LALLS

of chain scission and subsequent chain transfer by hydrogen to produce a methyl end group. During this process, there is an increase in the weight average molecular weight and a broadening of the molecular weight distribution. Linking reactions have undoubtedly occurred. The "Y" branch concentration has approximately doubled for this 2.0 Mrad irradiation of Marlex 6003 in the solid state, but it may be only one of several possible linking reactions. In this instance, the weight average molecular weight changes for Marlex 6003 were not so dramatic as observed following a similar irradiation of NBS 1475. Even so, the increase in the weight average molecular weight and the broadening of the molecular weight distribution strongly indicate that linking reactions are a very important part of the radiation chemistry of polyethylene. Once again, only the linking reactions leading to "Y" branches can be positively indentified.

The formation of "Y" branches is not sufficient to account for the loss in terminal vinyl end groups. It is interesting to compare these results to those obtained after a 4.0 Mrad irradiation in air. In the latter experiment, oxygen has an opportunity to compete with the terminal vinyl groups for primary and secondary radicals. No long chain branches could be detected after irradiation of Marlex 6003 in the presence of air. Chain scission obviously occurs as indicated by the substantial increase in saturated end group concentration and the substantial reduction in weight average molecular weight. Correspondingly, the molecular weight distribution narrows as expected for random chain cleavage. There is no clear evidence for chain-linking reactions. The main chain hydroperoxide group concentration is quite high indicating how effectively oxygen was able to compete for the secondary radicals. It is interesting that there is a reduction in vinyl end group concentration without an increase in the "Y" branch content. The role of terminal vinyl groups in the free radical process may be more complex and not limited to being only a free radical scavenger resulting in linking reactions.

The experiment involving thermal degradation/melt irradiation of NBS 1475 at 225°C in vacuum was also performed on Marlex 6003 with similar results. No evidence for "H-link" formation was obtained. Table VIII contains the structural changes as measured by NMR quantitatively (a) after heating Marlex 6003 for 24 hours in vacuum at 225°C and (b) after a subsequent 3.0 Mrad irradiation. Thermal degradation in vacuum leads principally to the same species that are observed upon irradiation of polyethylene, that is, saturated end groups, vinyl end groups, internal double bonds and long chain branches. In addition, there is an increase in weight average molecular weight and a broadening of the molecular weight distribution. A subsequent irradiation leads to a reduction in the terminal vinyl content, an increase in long chain branching but no perceptible change in the saturated end group content. There is a further increase in weight average molecular weight and a further broadening of the molecular weight distribution. Once again, there is a greater reduction in terminal vinyl group content than there is an increase in long chain branching. It is reasonable that the terminal vinyl groups can react with secondary radicals to form a long chain branch but this is, once again, apparently not the only outlet for terminal vinyl end

TABLE VIII

Structural Changes in Marlex 6003 Following Thermal Degradation in a Vacuum at 225°C for 24 hours and a Subsequent 3.0 Mrad Irradiation in the Melt at 225°C

Structural Unit	Number of Units per 10,000 Carbons: After 24 Hrs at 225°C	Number of Units per 10,000 Carbons: After 24 Hrs at 225°C + 3.0 Mrad Irrad.
Saturated End Groups[1]	18.6	18.6
Vinyl End Groups[1]	14.8	9.4
"Y" Branches	1.9	3.6
Trans Double Bonds	1.6	1.9
Cis Double Bonds	2.0	1.8
Butyl Branches	2.3	3.1
Hydroperoxide Groups	1.9	1.9
Carbonyl Groups	N.D.[2]	N.D.
$M_w \times 10^{-3}$	166[3]	192[3]
$M_n \times 10^{-3}$	13	15
M_w/M_n	12	13

[1]Concentrations determined from relative areas. No designation indicates peak heights were used.

[2]N.D. = None Detected

[3]Measured by GPC

groups. No evidence that would suggest the presence of "H"-links was present in the ^{13}C NMR spectrum of this sample.

From this study of the irradiation of polyethylenes up to the gel dose, it appears that "H"-link formation is less significant than the formation of long chain branches, cis and trans double bonds and chain scission. This does not infer that "H" link formation is unimportant. It is likely the event that leads to gel formation. The formation of star types of polyethylene molecules prior to any significant "H" linking would only mean that fewer "H" links are required to be effective in producing a network structure.

It is interesting to note that the increase in the content of saturated end groups occurred only with the room temperature solid state irradiations of polyethylene. No change in the methyl content was observed following any of the melt irradiations. There were three opportunities: the 1.0 Mrad melt irradiation of NBS 1475 following thermal degradation, the 3.0 Mrad melt irradiation of Marlex 6003 following thermal degradation and the 100 Mrad melt irradiation of n-hexatriacontane.

It is disappointing that ^{13}C NMR of polymer solutions do not reveal more about the structure of gelled polyethylene networks. In a related study, D. L. VanderHart of the National Bureau of Standards has used cross polarization-magic angle spinning (CP-MAS) to investigate polyethylenes radiated up to 200 Mrad at room temperature in the solid state(14).

5. Irradiation of Polyethylene - NMR of the Solid State

VanderHart examined three different polyethylenes in his solid state NMR study(14): NBS 1475, an ethylene-1-butene copolymer having 2.7 branches per 1000 backbone carbons and an ultra high molecular weight polyethylene (UHMWPE) from Hercules with an M_w exceeding 2 x 10^6. Each sample was irradiated in a vacuum at 25 and 167 Mrad using a ^{60}Co gamma source. The UHMPE was also subjected to electron beam irradiation in air at 100 and 200 Mrad. All the experiments were performed at ambient temperature conditions.

The most notable effect upon irradiation was a sizeable decrease in the observed $T_{1\rho}{}^{H}$s. Only nominal changes in crystallinity were observed. No resonances could be detected in the 38-41 ppm range for the methine resonances from either "H" links or "Y" branches. Only resonances corresponding to cis and trans double bonds and saturated end groups were observed following irradiation. The yield for cis and trans double bonds increased with radiation dose but not in a linear manner. There was no apparent dependence upon either density or molecular weight. The methyl group concentration was observed also to increase with increasing radiation dosage suggesting that chain scission was a product of the radiation process. The methyl group partitioning between phases favored the amorphous phase with a concentration of approximately twice that of the crystalline phase.

VanderHart monitored the terminal vinyl group behavior as well and observed it to diminish with increasing radiation dose. There was the suggestion that both cis and trans double bonds were present in the amorphous phase while the crystalline phase contained predominantly trans double bonds. No appreciable differences between vinylene concentrations between phases was detected in the UHMWPE sample. The resolution was insufficient to establish this relationship in the other two samples.

The CP-MAS observation by VanderHart that trans double bonds are found principally in the crystalline phase is consistent with the conclusion of previous investigators(15)(16). It also offers an explanation for the result obtained from solution NMR on the NBS 1475 sample irradiated to 8.0 Mrad. The only soluble portion of the sample was an essentially linear polymer with no terminal unsaturation and a considerable quantity of trans double bonds. It is possible that this portion of the polymer originated from the crystalline region and had undergone essentially no linking reactions.

It is not surprising that the CP-MAS experiment was unable to detect either the "Y" branch or the "H" link. The solution NMR work indicated the concentrations of these species to be sufficiently low to offer detection difficulties even under the best of circumstances. With the linewidths presently obtainable in CP-MAS studies, it is unlikely that these resonances will be visible.

6. Summary

It is gratifying that similar observations could be made following independent solution and solid state NMR studies of the radiation of polyethylene. Sensitivity is a problem in both studies as the species formed are in a concentration range of only a few units per 10,000 carbons. Line widths offer difficulties as well as it is highly likely that the solvent swollen network structures yield exceedingly broad resonances, thus limiting the available sensitivity to concentrations that are orders of magnitude higher than required for a meaningful study. Inspite of these difficulties, considerable structural information is available from NMR studies, particularly if the samples are examined prior to the gel dose.

From both the solution and solid state NMR studies of the irradiation of polyethylene, it could be concluded that

(a) chain scission leading to methyl terminated chains occurs during irradiation in the solid state,

(b) terminal vinyl groups are consumed during solid state irradiations,

(c) trans internal double bonds are formed in the crystalline regions of polyethylene during solid state irradiations and

(d) both cis and trans internal double bonds are formed in the amorphous regions of polyethylene during solid state irradiations.

The independent solid state NMR study of the radiation of polyethylene revealed the following additional information:

(e) the distribution of internal double bonds produced upon irradiation was observed to be about the same between the amorphous and crystalline phases and

(f) approximately twice as many methyl groups were observed in the amorphous phase as opposed to the crystalline phase of an irradiated UHMWPE.

The NMR solution examination of polyethylenes irradiated to a point short of the gel dose revealed the following:

(g) long chain, "Y" types of branches are formed during both melt and solid state irradiations of polyethylene,

(h) the formation of saturated end groups with irradiation occurred primarily in the solid state with little or no increase in methyl content observed following melt irradiations,

(i) terminal vinyl groups are consumed in both melt and solid state irradiations of polyethylene,

(j) irradiations of polyethylene conducted in the presence of air lead to a dramatic reduction in the number of linking reactions and

(k) the structural units produced in polyethylene are similar whether the sample is thermally degraded in the absence of air or subjected to gamma irradiation.

With the continuing improvements in the sensitivity of the ^{13}C NMR technique, more structural information about the radiation induced structural changes will be forthcoming. The CP/MAS study in the solid state has the advantage that the crystalline and amorphous phases of polymers can be examined independently. Thus the partitioning of structures between amorphous and crystalline phases can be determined directly. Such information is lost in solution NMR studies. In both techniques, there is still a question of having sufficient sensitivity to detect structural moieties at the required very low concentration levels.

The most important conclusion arising from the NMR solution studies is that the importance of the "Y" branch had not been previously realized. The failure of both NMR techniques to detect the "H" link raises questions not only about the capabilities of the NMR techniques but also about the presence of the "H" link in irradiated polyethylenes. It is clear that it is not an important structural moiety prior to the gel dose. What happens after that is still speculative. The production of "Y" branches appears to quite reasonably follow from a reaction of terminal vinyl groups and secondary, main chain radicals(15). The fact that there is not a material balance between the consumption of vinyl end groups and the formation of "Y" branches raises another important question. There are apparently other "outlets" for the consumption of vinyl groups during irradiations of polyethylenes.

7. Acknowledgements

The author is indebted to Professor Joe Silverman of the University of Maryland and to Dr. Frederick J. Zoepfl not only for their enthusiasm for this work but for many long and thoughtful discussions. The author is also indebted to Professor Olgun Güven for his invitation to participate in the NATO conference on scission and linking reactions in polymers, which provided the opportunity to review and organize this presentation.

8. References

(1) J. C. Randall, "Polymer Sequence Determination: Carbon 13 NMR Method", Academic Press, New York, N.Y., 1977

(2) H. J. Harwood and W. M. Ritchey, Polymer Letters, 1964, 2, 601

(3) J. C. Randall and C. J. Ruff, Macromolecules, In Press

(4) J. C. Randall, Journal of Macromolecular Science - Reviews in Macromolecular Chemistry and Physics, In Press

(5) J. C. Randall, "Polymer Characterization by ESR and NMR", ACS Symposium Series No. 142, Edited by A. E. Woodward and F. A. Bovey, 1980

(6) T. C. Farrar and E. D. Becker, "Pulse and Fourier Transform NMR", Academic Press, New York, N. Y., 1969

(7) J. Schaefer and D. F. S. Natusch, Macromolecules, 5, 416, 1972

(8) J. Research of the National Bureau of Standards, 1971, 76A(2)

(9) B. J. Lyons, Polym. Prepr., Am. Chem. Soc. Div. Polym. Chem., 8, 672 (1967)

(10) J. C. Randall, F. J. Zoepfl and J. Silverman, "NMR and Macromolecules," J. C. Randall, Ed., ACS Symposium Series No. 247, 245 (1984)

(11) J. C. Randall, F. J. Zoepfl and J. Silverman, Makromol. Chem. Rapid Commun, 4, 149 (1983)

(12) R. L. Bennett, A. Keller, J. Stejkny and M. Murray, J. Polym. Sci., Polym. Chem. Ed., 14, 3027 (1976)

(13) F. A. Bovey, F. C. Schilling and H. N. Cheng, "Advances in Chemistry Series", No. 169, Edited by D. L. Allara and W. L. Hawkins, 1978, Chapter 11

(14) E. Perez and D.L. VanderHart, J. Polym. Sci., Polym. Phys. Ed., 26, 1979 (1988)

(15) G. N. Patel, J. Polym. Sci., Polym. Phys. Ed., 13, 351, (1975)

(16) L. Mandelkern in "The Radiation Chemistry of Macromolecules", Vol. 1, Chapter 13, Edited by M. Dole, Academic Press, New York, 1972.

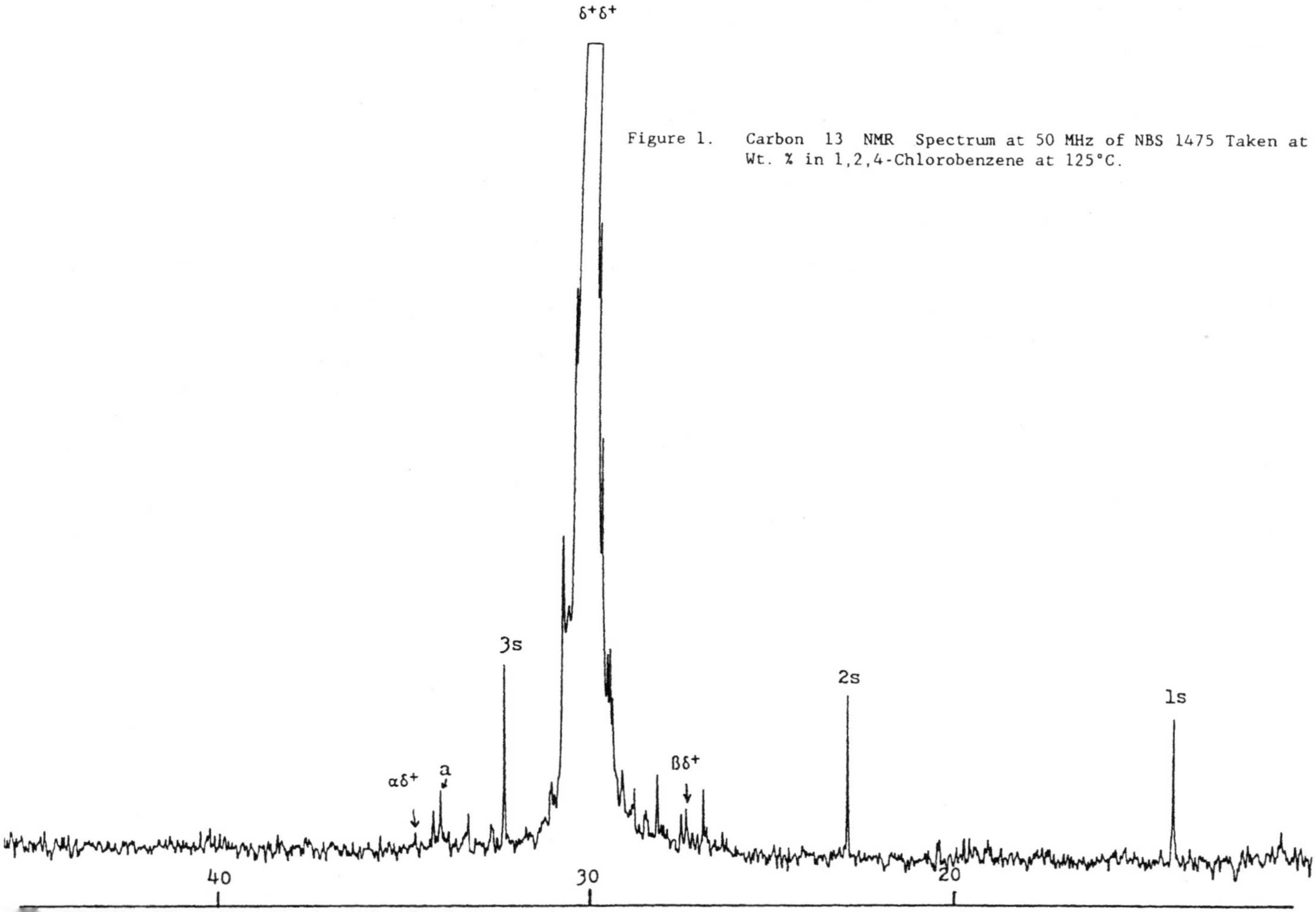

Figure 1. Carbon 13 NMR Spectrum at 50 MHz of NBS 1475 Taken at 15 Wt. % in 1,2,4-Chlorobenzene at 125°C.

Figure 2.
Carbon 13 NMR Spectra at 50 MHz of a Medium Density Polyethylene Taken Before and After Severe Melt Processing. NMR Solutions are at 15 Wt. % in 1,2,4-Trichlorobenzene at 125°C.

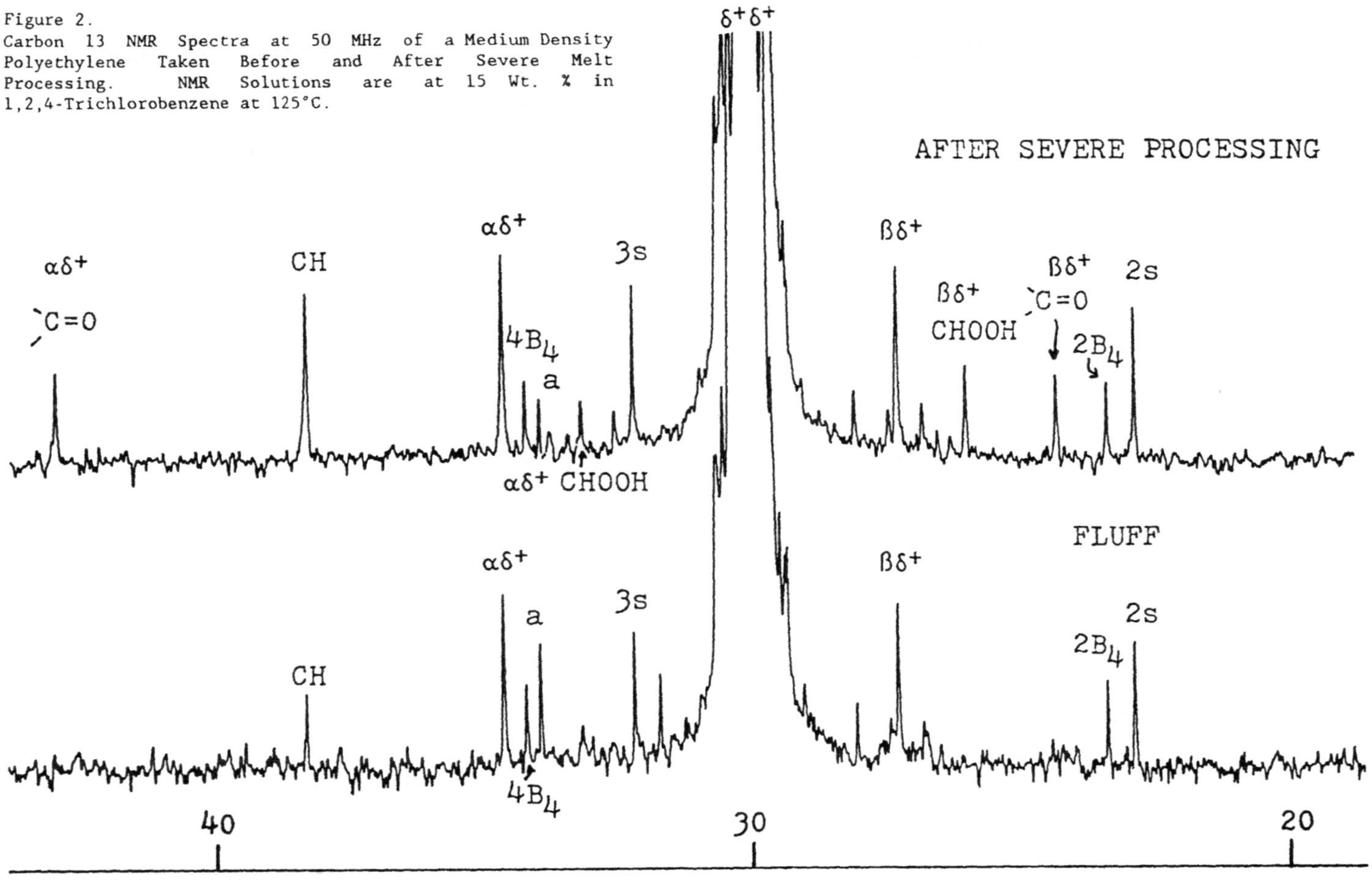

Figure 3.
Carbon 13 NMR Spectra at 50 MHz of NBS 1475 Taken Before and After a 2.0 Mrad Irradiation in the Solid State. NMR Solutions are at 15 Wt. % in 1,2,4-Trichlorobenzene at 125°C.

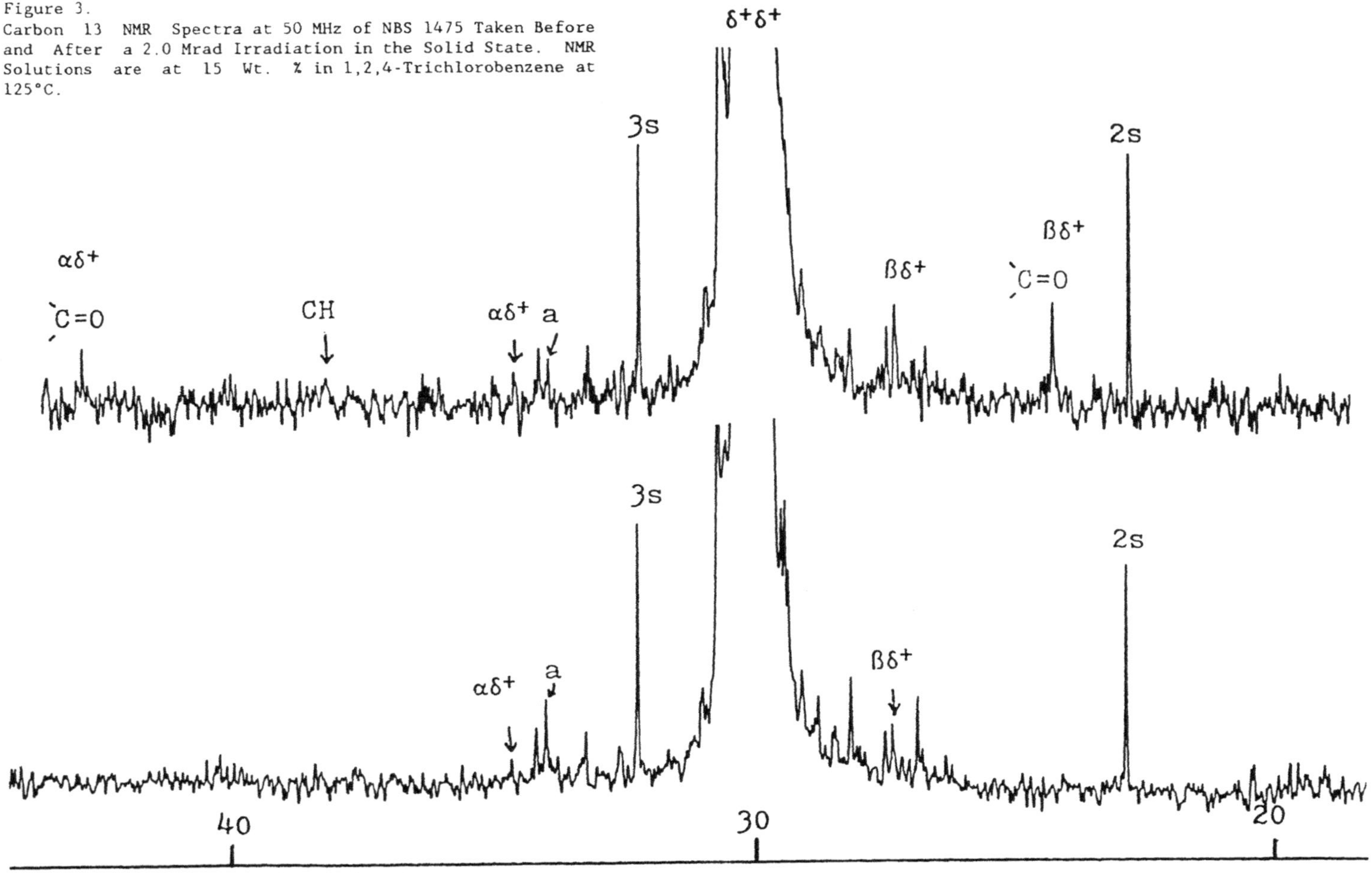

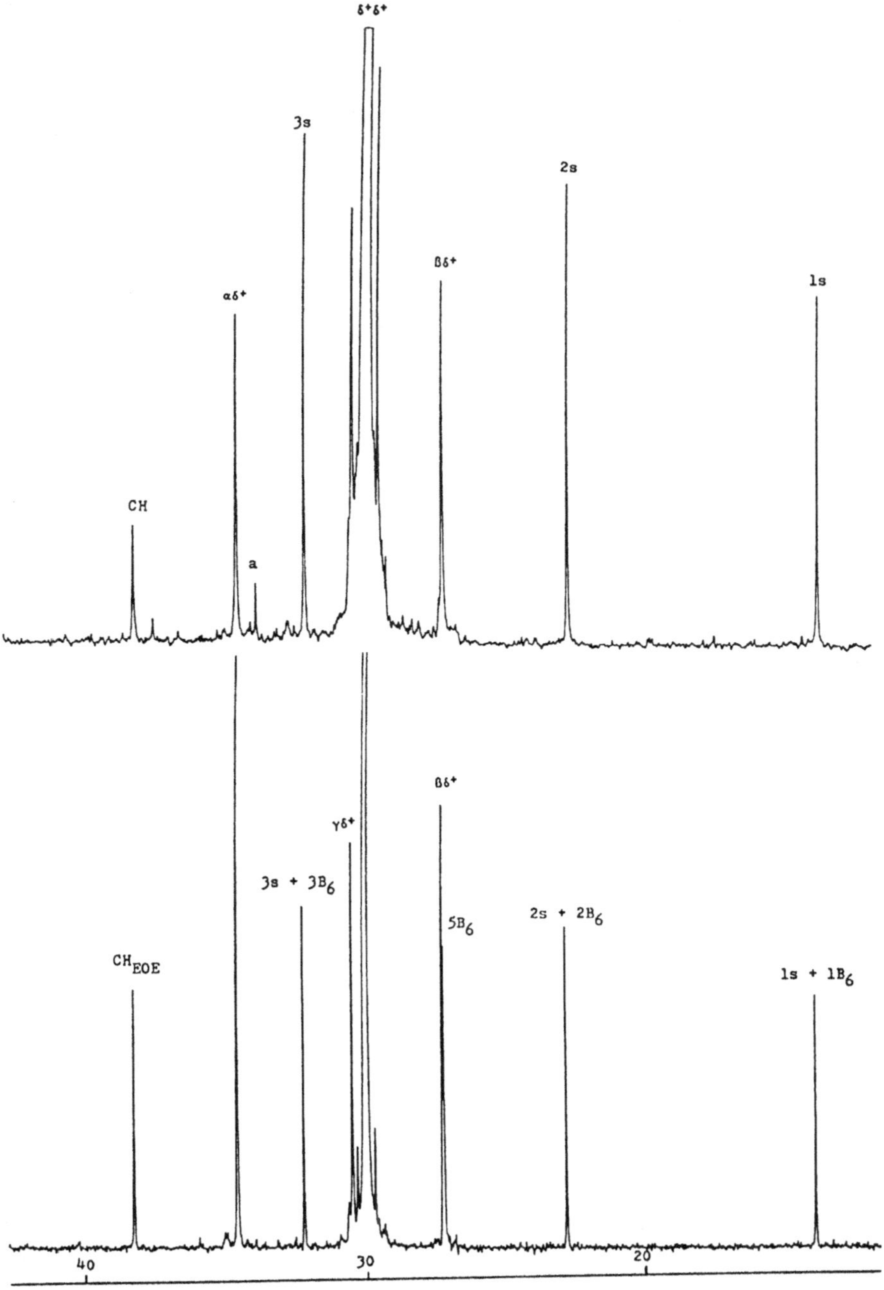

Figure 4. Carbon 13 NMR Spectra at 50 MHz of (lower) an Ethylene-1-Octene Copolymer and (upper) NBS 1475 after Severe Thermal Degradation in a Vacuum at 225°C and a Subsequent 1.0 Mrad Irradiation at 225°C. NMR Solutions are at 15 Wt. % and 125°C.

Figure 5.
Carbon 13 NMR Spectra at 50 HHz of Marlex 6003 Taken Before and After a 2.0 Mrad Irradiation in the Solid State. NMR Solutions are at 15 Wt. % in 1,2,4-Trichlorobenzene at 125°C.

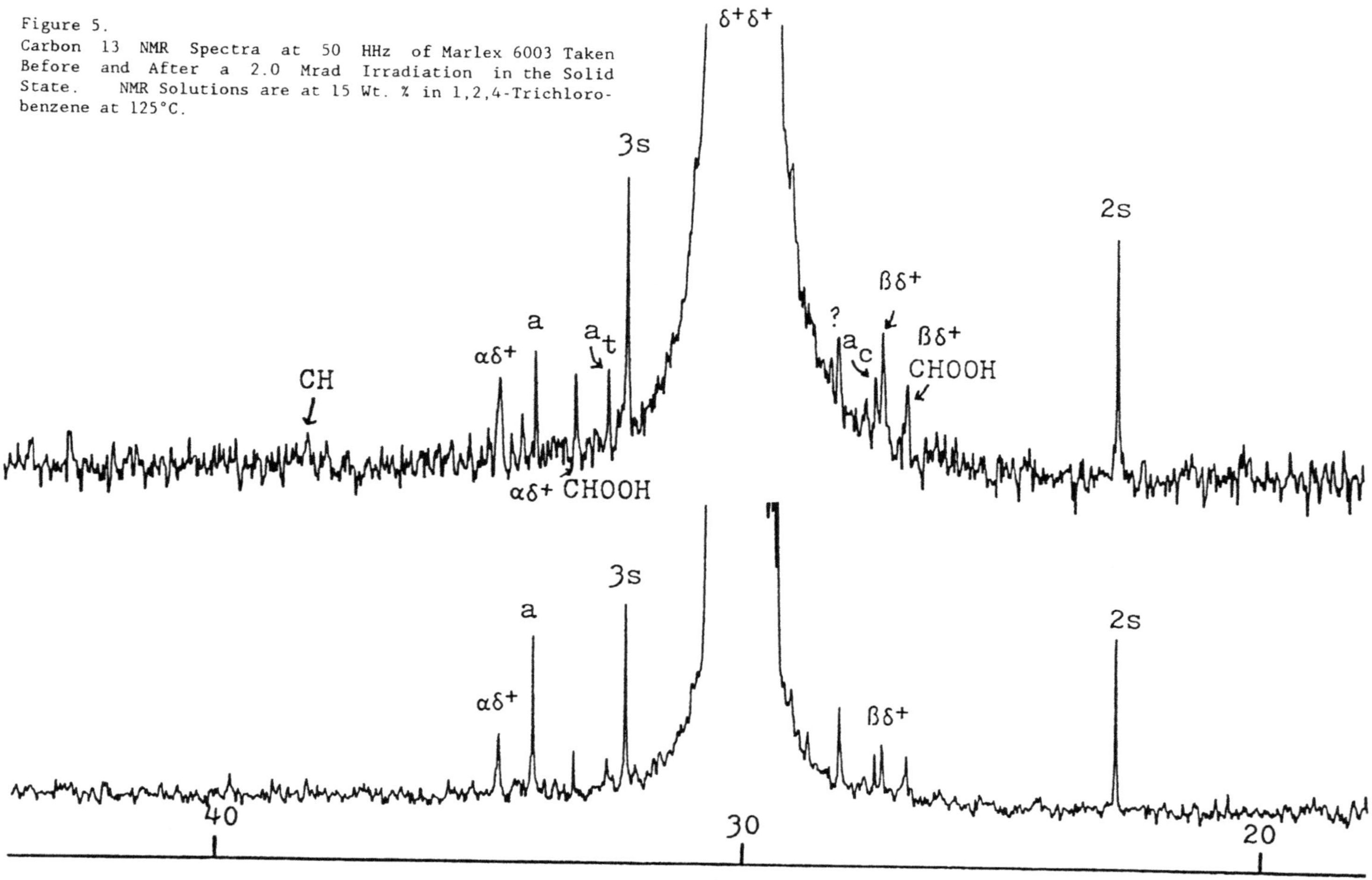

THE USE OF PULSED NMR TO DETERMINE NETWORK FORMATION BY CROSSLINKING AND ENTANGLEMENT

A. Charlesby,
Silver Spring, Watchfield,
Swindon SN6 8TF, UK

INTRODUCTION

Although nuclear magnetic resonance has been used for many years as a powerful technique for chemical analysis, its use for the determination of arrangement and mobility of high molecular weight systems is far from familiar. Yet it promises a powerful method of investigating such matters as molecular weight, mobility in a solvent, network formation by crosslinks or entanglements and other physical properties without any detailed knowledge of chemical structure . In what follows we shall be primarily concerned with the spin-spin relaxation curve (T_2) using a pulsed NMR technique

The spin-spin relaxation curve (free induction decay, F.I.D) can consist of components representing the rigid structure (crystalline or glassy), the mobile molecules in a flexible network, and the individual free molecules. At first we consider the relaxation curve for polymer molecules of low molecular weight (typically less than about 10^4) and in a mobile state, either because of a temperature well above the crystalline or glassy state, or because of a solvent. For the simplest case (no solvent) , the curve is usually a simple exponential decay , characterized by a time T_{2L}; $\exp(-t / T_{2L})$. T_{2L} is found to depend on average molecular weight M approximately

$$T^2_{2L} \propto M^{-1}$$

Since for such polymers its viscosity η varies approximately as M, it follows that T_2^{-1} measures the mobility of the polymer molecule. This is best shown with polyisobutylene, a polymer which suffers main chain scission in direct proportion to the radiation dose. In a series of samples of diminishing molecular weight, the product $T_{2L} M_n^{0.5}$ remains constant (at 150 °C) with a scatter of a few percent over a range in M_n from 5200 to 88.

O. Güven (ed.), Crosslinking and Scission in Polymers, 83–90.

Table I. Value of $T_{2L}\ M_n^{0.5}$ over a very wide range of molecular weight of polyisobutylene (< 10000)

r (Mrad)	0	2.16	7.5	32.8	78	140	155	290	427	967
M_n (calculated)	5200	4603	3586	1752	915	553	505	283	195	88
$10^3\ T_{2L}$ (observed)	52	62	72	96	125	158	168	210	270	400
$T_{2L}\ M_n^{0.5}$	3.75	4.21	4.31	4.02	3.8	3.72	3.77	3.53	3.77	3.75

The effect of such parameters as temperature, branching and solvent on ω S(2,2L) $M^{0.5}$ may be compared with their effect on viscosity.

For flexible polymers crosslinked to give a partially soluble network, the spin -spin relaxation curve is no longer a simple exponential, but can be represented by the sum of two exponentials

$$A(t) = f\ \exp(-t / T_{2L}) + (1 - f) \exp(-t / T_{2S})$$

and for initially low molecular weight polymer, the parameter f is found to correspond to the soluble fraction s as determined by the usual solvent technique. The parameter T_{2S} appears to be determined by M_c the average molecular weight between successive links in the network fraction, while T_{2L} refers to the mobility of the residual free molecules , obviously modified by their dispersion within the polymer network. Thus pulsed NMR appears to offer a rapid, simple and non-destructive method of measuring crosslinking and network formation. The value of T_{2S} is usually of the order of 0.1 to 10 ms, while T_{2L} is often far longer , sometimes about 100 to 1000 ms, so the two components can be readily distinguished. Since crosslinks are permanent chemical bonds, the sol and gel (network) fractions are independent of temperature, the quantities f and 1-f do not vary with temperature, although T_{2L} and T_{2S} do so vary since they depend on molecular mobility.

The importance of the limiting molecular weight (typically about 10^4, depending on factors such as temperature) is because above this value, the spin-spin relaxation curve shows two components f and 1-f (T_{2L} and T_{2S}) even for a polymer which is not crosslinked, and is completely soluble. The explanation is that chains are entangled on a effective temporary basis, but

long enough to appear as effective crosslinks during the very short time span needed for the NMR pulse measurement. Thus pulsed NMR measures crosslinks + entanglements and in the absence of the former, a temporary network can still appear to be formed at a sufficiently high concentration of entanglements, namely an average of at least one entangled unit per weight average molecule. For some polymers this occurs at room temperature at a molecular weight of about 10^4, i.e. there are $6x10^{19}$ entangled units per g of polymer.
The presence of entanglements as well as permanent crosslinks to form a network can be represented by the sum of two radiation doses r_e and r,, where r_e is a virtual dose which would produce a number of crosslinks equivalent to the entanglements, r_e would of course vary with temperature.
The presence of a network in a partially cross-linked system is often determined by the insoluble (gel) fraction, and may follow a theoretical relation (Charlesby - Pinner plot). Fig. 1 shows such an experimental plot of sol fraction s (=1-network) which first starts to diminish at a gelation dose r_{gel}. The curve follows the calculated Charlesby- Pinner relation approximately, as does the parameter f (representing the long term component T_{2L}) provided that for NMR the dose is not taken as r, but r + r_e . Here r_e = 6 Mrad is a virtual dose , which would be needed to produce as many additional crosslinks as there are entanglements per molecule . Under the conditions of the experiment this would be as M_c value of about 40000 .
Although entangled and crosslinked units appear to behave similarly as far as network formation is concerned, at least for a period of less than 1s, they can be readily distinguished since the crosslinks are permanent bonds , while the density of entangled units decrease rapidly with temperature, and the limiting molecular weight rises rapidly above 10^4 . Thus the value of 1-f (due to both crosslinks and entangled units forming a network) decreases rapidly with rising temperature, but reaches a limiting value when all the

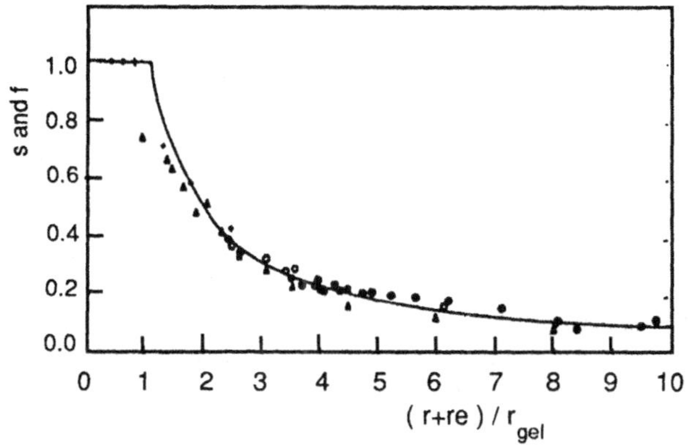

Fig. 1 The change of sol fraction with dose

effective entangled units have disappeared, and only the crosslinks remain . In this manner one can measure not only the number of entanglements, but also their variation with temperature. In fact it may be that the number of entanglements remains constant, but their lifetime is diminished, and new entanglements replace them. Only the entanglements of lifetime longer than that needed for the pulsed NMR are effective for this technique. Thus it appears that we can measure entanglement concentrations, their dependence on temperature and their average lifetime.
A series of three polydimethyl siloxanes (PDMS) show f (free molecule fraction, Fig. 2, equivalent to soluble fraction) of 0.1, 0.3 and 0.65 at T=220 K, but at higher temperatures these fraction diminish as the number of effective entanglements diminishes. At about 340 K the 26000 sample reaches f=1, i.e. there is just one entangled unit per molecule, M_c=26000. For the 67000 sample the point at which f=1 occurs at about 490 K i.e. M_c = 67000, 9 x 10^{19} entangled units per g (4.5 x 10^{19} entanglements) . By a more detailed analysis one can evaluate entanglement density over a range of temperatures.

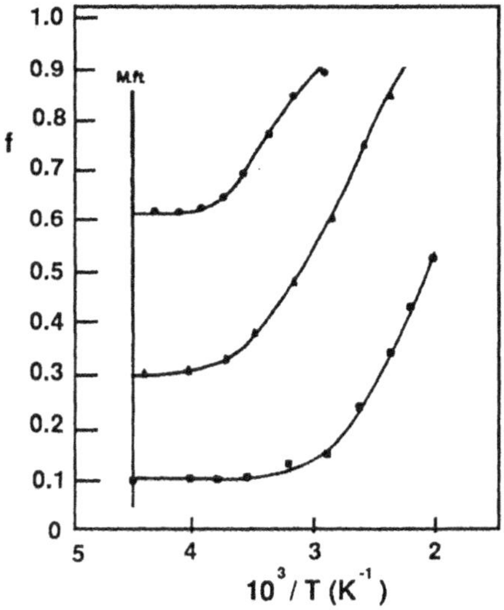

Fig. 2 f values for the three whole polymer samples of PDMS plotted agaginst reciprocal temperature (●) Mw ≈ 26,000. (▲) 67,000 and (■) 110.000.

Although the word entanglement is used in this content, one is really inferring some form of association between neighbouring molecules, best envisaged as an entanglement.
As might be expected the introduction of a solvent greatly increases molecular mobility and diminishes the number of effective entanglements detected by NMR. This can be clearly demonstrated by the reduction of (1-f) , equivalent to that obtained by a temperature rise. If it is assumed that

polymer mobility is what these measurements involve, an equivalence can be posed between temperature rise, and increase in volume due to solvent. It can therefore be shown that the mobility is simply a function of the free volume obtained by either of these two means, and since this NMR measurement parallels viscosity, it appears that viscosity also depends on the free volume (as well as molecular weight).

The above refers to molecules in a highly flexible state, either as in a highly elastic network, or as free molecules, these being characterized by the two exponential decays with characteristic times T_{2L} and $T_{2S,}$ with typical times of the order of ms or even sec. For polymer of which a fraction may appear rigid (as in crystals or a glassy system), a third component appears denoted by $T_{2C,}$ of the order of ms. Moreover unlike T_{2L} and $T_{2S,}$ this free induction decay is no longer exponential (Lorentzian) but is approximately Gaussian, i.e. $\exp(-t^2 /T^2_{2C})$ or more generally $\exp (- t^a / T^a_{2C})$. Here T_{2C} is far smaller than T_{2S} , of the order of 10 ms , so that it can only be picked up from the log (intensity) /log time curve for the shortest times, and it may appear less certain. The transition from T_{2C} to T_{2S} can be readily seen, and it becomes possible to use its measurement to determine crystallization rates. The results thus obtained with pulsed NMR agree with those derived from thermal analyses, but cover a much wider range of times , and the measurements can be made very quickly.

Many attempts have been made to deduce the effect of additive acting as stabilizers for irradiated polymer, and their effect can be readily followed using the FID curves to follow the effect of irradiation for a series of samples They have also been used to determine the effect of fillers such as carbon particles in rubber.

In one or two polymers subjected to this type of NMR analysis, and well above the crystallization or glass transition temperatures (T_c or T_g), a third relaxation component has been noted, with a T_m value between T_{2S} and T_{2C} . Its amplitude is usually very small, but its origin is as yet unknown, (and indeed its occurence is not fully confirmed. There is of course room (or even possibly a requirement) for an additional component corresponding to the motion of the fractions of a chain, tied into the network at only one point, and for which M_c has no meaning. It may also originate from some other form of structure, due to H bonding for example, since it does not occur in most polymers, and is in any case small, or it may occur in chains at the surface of these more rigid volumes, or within them, at regions of local disorder. Some measurements indicate that it only appears above the gel point.

The influence of solvent and of temperature on these entanglements has been studied in the case of monodispersed polyisoprene, for polymer of molecular weight 7200, where the number of entanglements is insufficient to form a network at room temperature, only a T_{2L} component is seen and this varies with solvent concentration and with temperature in a very characteristic manner, showing that (apart from molecular weight), it is dependent primarily on free volume, whether this be afforded by the solvent or an increase in temperature. For higher molecular weight polyisoprene , both T_{2L} and T_{2S} components are present, revealing the existence of an entangled network, with M_c in the neighbourhood of 40 -50000. Polyisobutylene is a polymer which scissions under radiation, and the transition from a high molecular weight, highly entangled system, with substantially only a T_{2S} component due to network , to a low molecular weight, unentangled (i.e.-non network) system, with only a T_{2L} component, can be followed quantitatively (Fig. 3 .

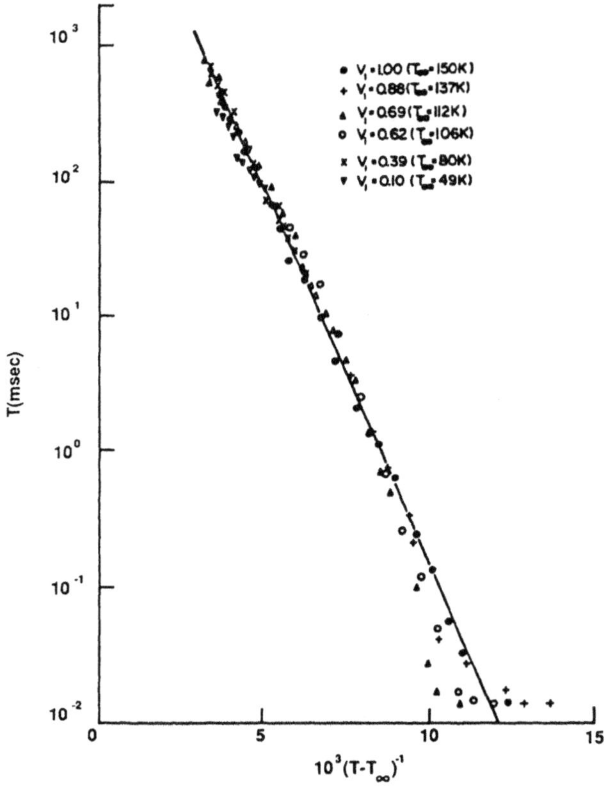

Fig. 3 Relationship between spin-spin relaxation time (T_{2L}) and temperature T for monodisperse polyisoprene Mw = 7200 at varios concentrations V1'

Polyisobutylene at 144 °C, pulsed NMR pattern T_2 relaxation for doses from 0 to 40 Mrad, with a change in molecular weight, the transition from only T_{2S} to partial $T_{2S} + T_{2L}$ and finally T_{2L} component is clearly shown.
The pulsed NMR method has been used to follow crystallization rates over long periods, and at any desired temperature. The results agree with those derived by more direct means, but thanks to its flexibility, the NMR method allows the transition to be followed over a much wider range.
The changes due to the irradiation of polymer aqueous solution can also be traced using a D_2O solution. The curves show only a T_{2L} component, as the solvent reduces the lifetime of any entanglement below that seen by NMR.
The technique can also be used to follow polymerization rates at various temperatures showing a rapid drop in T_{2L} and f after a certain time (which depends on temperature) as the monomer polymerizes to give polymer which can then entangle to form a network.
The use of pulsed NMR has also been extended to study radiation effect in some biological systems, such as Bovine Serum Albumin, in water. This shows the expected simple exponential decay, whose slope is greatly increased with a relatively small dose. The presence of a very small amount of protector such as thiourea almost completely eliminates this effect of radiation-far more than might be expected if it only intervenes in protecting the biomolecule against the radiation products of the water.

REFERENCES

(a) General

A. Charlesby; Rad. Phys. and Chem. ,14, 919 (1979)
Tihany Symposium, Budapest, 843 (1983)
A. Charlesby; Jnl of Radioanalytical and Nuclear Chem. Articles 101 (2) 401 (1986)
A. Charlesby, R Folland ; Rad .Phys. and Chem. , 15 , 393 (1980)
Rad. Res. Conf., (Tokyo) 336 (1979)
A. Charlesby (Polym. and Biological Macromol.) Polymer J.,19 (5) 649(1987)

(b) Polyethylene

R Fallow , A . Charlesby, Europ. Polym. J., 15 , 953 (1979)

J. Polymer Sci. Let. Ed ., 16 , 339 (1978)
B. J. Bridges, A .Charlesby, R. Folland , Proc. Roy. Soc. (London) A (1979) 367, 343
I. Kamel , A .Charlesby; J Polym . Sci, (Phys. Ed.) 19 ,803 (198)

A .Charlesby, R. Raefer, R. Folland, Rad. Phys. Chem, 11, 83 (1978)
(c) Poly dimethyl Silicones

R. Folland, A .Charlesby Int . J. Rad . Phys. Chem ., 8 ,555 (1976)
R. Folland, A. Charlesby Rad . Phys. Chem. 10, 61(1977)
J. Poly. Sci .,16, 1041 (1978)

(d) Poly Isobutylene

A .Charlesby , J . Steven, Int . J. Rad. Phys . Chem ., 8 , 585 (1976)
A. Charlesby , B J. Bridges, Rad. Phys. Chem., 19, 155 (1982)

(e) Poly Styrene

A. Charlesby , E. M. Joroszkiewiez , Europ. Pol. J. ,21 , 55 (1985)

(f) Poly cis Isoprene

R. Folland , A. Charlesby, Polymer, 20 , 207, 211 (1979)
A. Charlesby, B.J. Bridges, Europ . Pol. J., 17, 645 (1981)

METHODS FOR STUDYING THE EFFECTS OF HIGH-ENERGY IRRADIATION ON THE MECHANICAL AND ELECTRIC PROPERTIES OF POLYMERS

Peter Hedvig
Vajda Peter u 11
Budapest 1089 Hungary

INTRODUCTION

High energy electron-, X-ray- or gamma-irradiation is known to result in degradation and/or crosslinking of polymers [1]. Crosslinking usually dominates when the polymer is irradiated in the absence of oxygen while degradation is overwhelming when oxygen is present.

The aim of this paper is to discuss how to measure the effects of high energy irradiation on the main mechanical and electric properties of polymers.

In the following sections the most important properties will be defined and the methods of their measurement discussed with illustrative experimental examples.

MECHANICAL PROPERTIES

The response of a material to a mechanical load is expressed in terms of a response function which describes how the system would reach thermodynamical equilibrium after being subjected to mechanical action . This action may be a constant strain , ε_0 , which results in a decaying stress, σ (t). This stress may be expressed in terms of the time-dependent stress-relaxation response function ϕ_σ (t) as

$$\sigma(t) = \sigma(\infty) + [\sigma(0) - \sigma(\infty)]\,\phi_\sigma(t) \qquad (1)$$

where $\sigma(\infty)$ is the stress at $t \longrightarrow \infty$; $\sigma(0)$ is that just after straining; ϕ_σ (t) is dimensionless and is normalized as $\phi_\sigma(0) = 1$; $\phi_\sigma(\infty) = 0$. This function is independent on strain in the so called "rheologically linear" range which is for elastomers ε_0=100-400% , for uncrosslinked polymers up to about 50 % in

O. Güven (ed.), Crosslinking and Scission in Polymers, 91–128.

certain temperature ranges, for semicrystalline polymers up to about 2 %. Amorphous polymers below their glass-transition temperature are essentially rheologically nonlinear systems.
Equation (1) describes the response of the material to a step-function strain. When the strain is time-dependent, ε (t), the response is expressed as

$$\sigma(t) = E(0)\,\varepsilon(t) - [E(0) - E(\infty)] \int_{t'=0}^{t} (d\phi_\sigma(t-t')/dt')\,\varepsilon(t')\,dt' \qquad (2)$$

where constants E (0) and E (∞) are referred to as limiting moduli. This expression can be derived by assuming that the response to any change of strain is independent on previous strains, i.e. the system has completely forgotten its history. (Boltzmann's superposition principle, see e.g. in Ref. [2].) This is another important condition of rheological linearity besides the independency of ϕ_σ on the strain.

Nonlinear cases can also be handled, but the expression is more complicated than Equ .(2). In the case of rheological linearity by Equ. (2) the response (stress) to any strain function can be calculated provided that the response to a step-function is known . When, for example , the strain is a harmonic (sinusoidal) function with an amplitude ε_0 and angular frequency ω

$$\varepsilon(t) = \varepsilon_0 \exp(i\omega t) \qquad (3)$$

the response (stress) is expressed by using Equ. (2) as

$$\sigma(t) = \sigma(0)\exp(i\omega t) + [\sigma(0) - \sigma(\infty)] \int_{t'=0}^{t} \left[(-d\phi_\sigma(t-t')/dt')\right] \exp(i\omega t')dt' \qquad (4)$$

where $\sigma(0) = \varepsilon_0 E(0)$; $\sigma(\infty) = \varepsilon_0 E(\infty)$

The stress-response to a harmonic sinusoidal strain is, therefore, a Fourier transform of the negative derivative of the function ϕ_σ (t).
Similary, when a constant stress σ_0 is applied to the polymer the strain is expressed as

$$\varepsilon(t) = \varepsilon(0) + [\varepsilon(\infty) - \varepsilon(0)]\,\phi_\varepsilon(t) + (t/\eta)\,\sigma_0 \qquad (5)$$

where ϕ_ε (t) is the creep-response function, which is dimensionless and is normalized as $\phi_\varepsilon (0) = 0$; $\phi_\varepsilon (\infty) = 1$; the last term of Equ. 5 describes irreversible viscous flow, η being the viscosity.

In case of time-dependent stress the strain-response is expressed in the "linear "case as

$$\varepsilon (t) = D (0) \sigma (t) - [D (\infty) - D (0)] \int_{t'=0}^{t} [d\phi_\varepsilon (t-t')/dt'] \sigma(t')dt' \tag{6}$$

where constants D (0) and D (∞) are referred to as limiting tensile compliances extrapolated to zero and infinite time respectively.

Modulus and compliance

From the above outlined picture it is clear that there are two basic experiments for characterizing the mechanical behaviour of a polymeric material. One is to keep the strain constant and measure the stress as a function of time (stress relaxation experiment), the other is to keep the stress constant and measure the strain as a function at the temperature (creep experiment). The characteristic parameters are the stress relaxation modulus.

$$E (t) = \sigma (t) / \varepsilon_0 \tag{7}$$

and the creep compliance

$$D (t) = \varepsilon (t) / \sigma_0 \tag{8}$$

where ε_0 is the constant strain, σ_0 is the constant stress. The strain may be defined simply as the change of length Δl over the original length l_0, "engineering strain" or as

$$\varepsilon_t = \ln (1 + \Delta l/l_0) \tag{9}$$

"true strain",in which the change of the cross-sectional area during extension is taken into account assuming that the total volume is unchanged.

Similarly the engineering stress is defined as the force, F, over the cross-sectional area, A, and the "true stress" as

$$\sigma_t = F / A\,(1 + \Delta l / l_0) \qquad (10)$$

σ and E are expressed in MPa (N / mm^2) units.
In torsion the deformation is measured in radians and instead of force a torque is applied. The torsional modulus can be expressed in terms of the extensional one as

$$\sigma = E / 2\,(1+\nu) \qquad (11)$$

where n is the Poisson's ratio i.e. the ratio of the strain measured in direction perpendicular to that of parallel with respect to the direction of stretching. For most polymers in extension n=0.3-0.5, in compression n= 0.5-0.7. n=0.5 corresponds to that case when the total volume is unchanged by deformation.
The torsional creep compliance is

$$J=2\,(1+n)\,D \qquad (12)$$

The stress relaxation modulus and creep compliance are connected by the following equations

$$\int_{t'=0}^{t} E\,(t-t')\,D\,(t')\,dt'=t$$
$$\int_{t'=0}^{t} D\,(t-t')\,E\,(t')\,dt'=t \qquad (13)$$

These integrals can be calculated numerically by the method of Hopkins and Hamming [3].

Thus, in the range of rheological linearity, it is sufficient to measure either stress relaxation modulus or creep compliance because they can be transformed into each other. As a rude approximation

$$E \approx 1/D\,;\quad G \approx 1/J$$

By application of harmonic load-(stress or strain)-function, as by Equ. (4), the transformation results in real- and imaginary-parts. For stress relaxation in extension e.g.

$$E'(\omega) = E_\infty + (E_0 - E_\infty)\, f'_\sigma(\omega)$$
$$E''(\omega) = (E_0 - E_\infty)\, f''_\sigma(\omega) \tag{14}$$

where the real part, E', is referred to as "storage-modulus", the imaginary one, E", as the "loss-modulus" f'_σ and f''_σ are the Fourier transforms of the negative derivative of the stress relaxation response function $\phi_\sigma(t)$, indexes 0, and ∞ refer to ω---> 0; ω--->∞ respectively. The compliances can be expressed (transformed) similarly. Since by harmonic loading the Poisson's ratio becomes also complex; the inter-relations between extensional and torsional moduli become

$$E' = 2G'(1+\nu') + 2G''\nu''$$
$$E'' = 2G''(1+\nu') - 2G'\nu'' \tag{15}$$

The mechanical loss tangents, defined as E"/E' or G"/G', are also used. (See e.g. in Ref. [2])

Ultimate mechanical properties

As was discussed in the previous section creep compliance or stress relaxation modulus are defined only within the limits of rheological linearity i.e. assuming that the creep response function is independent on strain, the stress relaxation response function is independent on stress and both response functions are independent on any previous actions ("history"). These requirements are not very often fulfilled. In many conventional quality control tests the samples are subjected to extreme loads much above linearity. In the usual stress-strain tests the strain is increased at constant rate and the stress is measured as a function of strain. At low strain-levels Equ. (2) may be used by substituting $\varepsilon(t) = \dot{\varepsilon}\, t$, where $\dot{\varepsilon}$ is the strain-rate . The resulting stress is obtained from Equ. (2) by integration by parts as

$$\sigma(t) = E(\infty)\dot{\varepsilon} t - [E(0) - E(\infty)]\dot{\varepsilon} \int_{t'=0}^{t} \phi_\sigma(t-t')dt' \tag{16}$$

At high strains the system becomes nonlinear because its structure changes. The stress-strain curve usually reaches maximum (yielding) then a plateau or quasi-plateau is formed until rupture . The main parameters describing the ultimate properties are the yield-stress (σ_Y) and-strain (ε_Y) and rupture-stress (σ_r) and-strain (ε_r) (See Fig.1). In usual uncrosslinked polymers in a certain temperature range the stress-strain curve reaches maximum after which the material flows with not much change of stress until rupture. By radiation induced oxidation the yield-stress is slightly increased and the rupture-strain is drastically reduced. For crosslinked polymers above their main transition temperature there is usually no yielding: the stress continues to increase with strain almost proportionally. From Equ. (16) it is clear that the stress-strain curve is governed by relaxation phenomena i.e. it depends on the temperature and on the strain-rate.

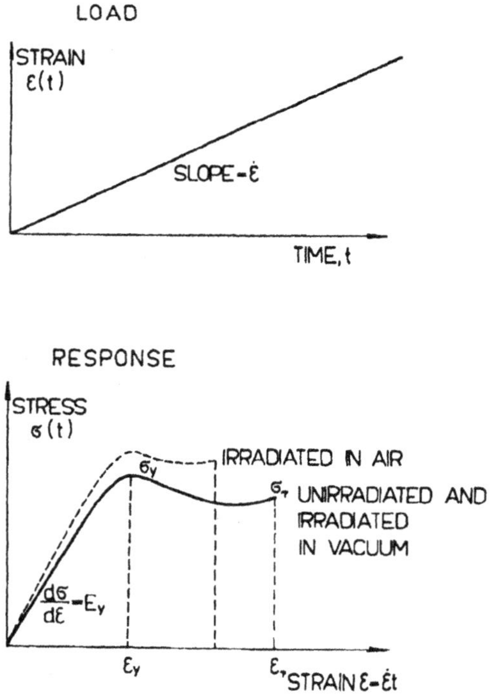

Fig. 1 Effect of irradiation on the stress-strain curve of a polymer. (Schematic)

The Young-modulus is defined as the slope of the stress-strain curve at small strains: (cf . Equ. 16):

$$E_\gamma = (d\sigma / d\varepsilon)_{t=to} = E(\infty) + [\, E(o) - E(\infty)\,]\, \phi_\sigma\, (t_o) \qquad (17)$$

This means, that the Young modulus determined from the stress-strain experiment is equivalent to the stress relaxation modulus determined from a step-function straining experiment at a constant time t_o.

Recovery

A very important property of most polymers is that their original shape is partially or fully recovered after certain mechanical and/or thermal loads. In practice it is important to know the load- (stress or strain) and temperature-ranges within which the response is fully recovered. Failure of recovery may be caused by viscous flow (usually at high temperatures), by crazeing or by structural changes at high load levels.
Recovery can be studied by applying a train of square load-pulses as indicated in Fig .2 and watch the recovery of deformation. Fig. 2 shows schematically the extreme cases. A completely anelastic material would not recover at all. An ideal elastic material would recover completely and promptly.

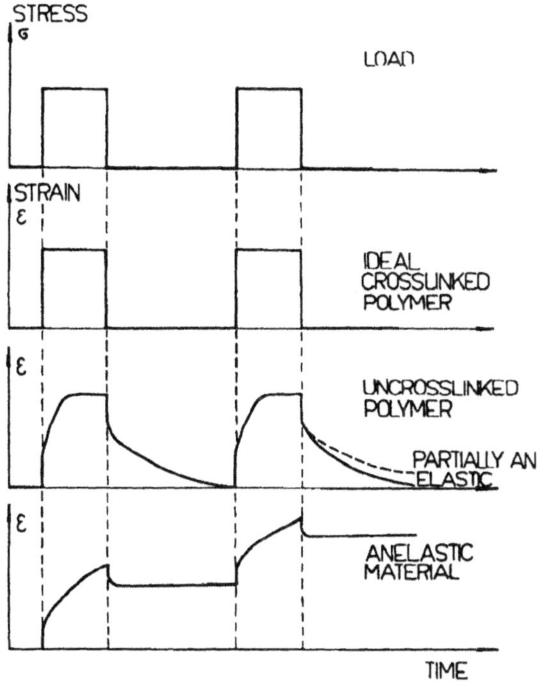

Fig.2 Response of different polymer types to square-shaped load pulses

An uncrosslinked polymer would recover partially and relatively slowly. Besides of using force-pulses one can use a train of strain-pulses and watch the recovery in-between them. Thus recovery may be studied after creep as well as after stress relaxation. Recovery may be described in terms of a response function $\phi_R(t)$ similarly to creep and stress relaxation. The time-dependent recovery of the strain after a certain pre-straining may be expressed as

$$\varepsilon(t) = \varepsilon(\infty) + [\varepsilon(0) - \varepsilon(\infty)]\,\phi_R(t) \tag{18}$$

where the recovery function is normalized as $\phi_R(o) = 1$, $\phi_R(\infty) = 0$ $\varepsilon(o)$ is the instantaneous deformation just after prestraining has been finished, $\varepsilon(\infty)$ is the unrecovered "anelastic" part of the deformation.

ELECTRICAL PROPERTIES

By applying an electric field to a polymer the response is current which arises from the displacement of charged particles within the material . There may be trapped "mono"-charges which would move by hopping from one side to another and finally to the electrodes and annihilate there. These result in irreversible current. In polymers almost always dipoles connected to the chains are found either because of the chemical construction of the material or by other effects as oxidation or diffusion of polar molecules from the environment.

Reversible chain-dipole polarization can be detected by measuring the current response to an electric field-step repeatedly till it is reversible. The response of the material to an electrical force field $\mathcal{E}$ can be treated similarly to the mechanical case if the polarization or the displacement function D is regarded as response. Similarly to the mechanical case by assuming linearity, i.e. that the response (polarization) to an external field is independent on previous actions and that the dielectric response function ϕ_d is independent on the applied field , one obtains [4]

$$D(t) = \varepsilon_v\, \varepsilon(o)\, \mathcal{E}(t) - \varepsilon_v\,[\varepsilon(\infty) - \varepsilon(o)] \int_{t'=0}^{t} [d\phi_d(t-t')/dt']\; \mathcal{E}(t')\, dt' \tag{19}$$

where ε_v is the permittivity of vacuum ($8.854\ 10^{-12}$ F/m), $\varepsilon(o)$ and $\varepsilon(\infty)$ are limiting permittivity values , ϕ_d is normalized as $\phi_d(o) = 0$; $\phi_d(\infty) = 1$

When $\mathcal{E}(t) = \mathcal{E}_o$ = constant i.e. step-function:

$$D(t) = \varepsilon_v \, \varepsilon(o) \, \mathcal{E}_o + \varepsilon_v \, [\varepsilon(\infty) - \varepsilon(o)] \, \mathcal{E}_o \, \phi_d(t) \qquad (20)$$

Equ. (19) may be expressed in terms of time-dependent permittivity which is defined as

$$\varepsilon(t) = D(t) \, / \, \varepsilon_v \, \mathcal{E}(t) \qquad (21)$$

resulting in

$$\varepsilon(t) = \varepsilon(o) - (\varepsilon(\infty) - \varepsilon(o) \, / \, \varepsilon_v \mathcal{E}(t)) \int_{t'=0}^{t} (d\,\phi_d(t-t') \, /dt') \, \mathcal{E}(t') \, dt' \qquad (22)$$

At constant external electric field strength the time-dependent dielectric permittivity is obtained from Equ. (22) as

$$\varepsilon(t) = \varepsilon(o) + [\varepsilon(\infty) - \varepsilon(o)] \, \phi_d(t) \qquad (23)$$

Equ. (23) is similar to the creep Equation (5), so the dielectric response function $\phi_d(t)$ is analogous to the mechanical creep response function $\phi_\varepsilon(t)$. In order to measure ϕ_d the electric charge should be measured as a function of time after the application of a voltage-(field)-step. This is technically not very easy, easier is to measure the current or current-density (current/area) which is expressed as

$$j(t) = [d\,D(t)/dt] + j(\infty) \qquad (24)$$

Comparison of Equ. (24) with Equ. (19) shows that the response function describing the current decay $j(t)-j(\infty)$ is the derivative of the dielectric response function (cf Equ. 20).

Thus, the dielectric response function may be measured by a "charge" experiment i.e., when the dielectric field is switched on suddenly and the current response is detected, and by a "discharge" experiment when the electric field is switched off and the discharge current is measured as a function of time.

By the "charge" experiment the current does not tend to zero: the value extrapolated to infinite time, $j(\infty)$, is referred to as being a result of "ohmic" conductivity involving continuous charge transfer in-between the electrodes.

The transient charge-and discharge-currents are mainly due to dipole polarization. Since these dipoles are in most cases bound to the polymer molecules the dielectric response detects the mobility of such parts of the molecule which are labelled by polar groups.

The "ohmic" conductivity may be a result of intrinsic charge activation or that of injection of charge carriers from the electrodes. High energy irradiation, evidently, creates a lot of "free" charge carriers which produce current when an external electric field is applied. If an electrical field is continuously on irradiation results in a significant increase of the current . When radiation is off, the current decreases. This is referred to as radiation induced conductivity and is illustrated on Fig. 3.

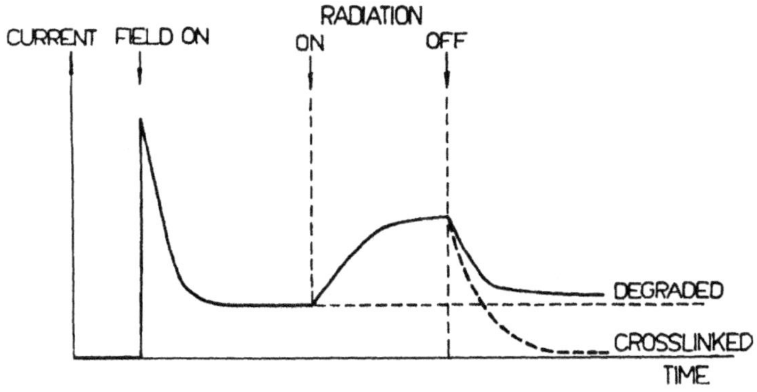

Fig.3 Response of a polymer to an electric field-step and to a high energy radiation pulse

By irradiation the conductivity increases to reach a certain level. When radiation is off, the conductivity decays to reach a value which is equal, lower, or higher to the initial "dark" conductivity-level indicating irreversible change of the structure of the material. When e.g. an uncured, or partially cured, resin is irradiated, the post-irradiation conductivity usually decreases because of crosslinking. When e.g. a noncrosslinkable polymer, as pure poly-vinylchloride (PVC) is irradiated, the post-irradiation conductivity usually increases because of partial conjugation of the chains and because of the presence of degradation products.

MULTIPLE TRANSITIONS AND MOLECULAR MOBILITIES

Polymers are known to exhibit several transition temperature-ranges where parts of their molecules become mobile. One of the lowest-temperature transition is connected with liberation of the rotation of the rotation of methyl groups connected to the main chain. Such transitions usually occur near to the temperature of liquid Helium. Liberation of the rotational or other types of motion of larger groups are observed at higher temperatures . As there are several types of motion of the polymer molecule, including the main-chain and side-groups, it is expected and found that polymers have many transition temperature-ranges which can be assigned to liberation of motion of different parts of the molecule.
Each transition has its own response function. The mechanical methods "mechanical spectroscopy" would detect all the transitions while dielectric spectroscopy detects only those in which motion of dipoles are involved.

Transitions in polymers may be characterized by their relaxation time (τ) distribution function

$$F(\ln\tau) = \sum_{i=1}^{n} c_i \left[d^i\phi\ (\ln t) \ /\ (d\ln t)^i \right]_{t=\tau} \qquad (25)$$

where ϕ stands for the mechanical or dielectric response function, the c_i coefficients are respectively: 0.434; $2.41\text{x}10^{-2}$; $5.02\text{x}10^{-3}$; $-3.35\text{x}10^{-4}$; $2.79\text{x}10^{-5}$; $-7.75\text{x}10^{-7}$. In practise n=3 is accurate enough. Thus by successive derivation of the response function the distribution can be obtained.

Fig. 4 shows an example when the response function has the form of

$$\phi(t) = \exp\left[-\ (t/\tau\ )^N \right] \qquad (26)$$

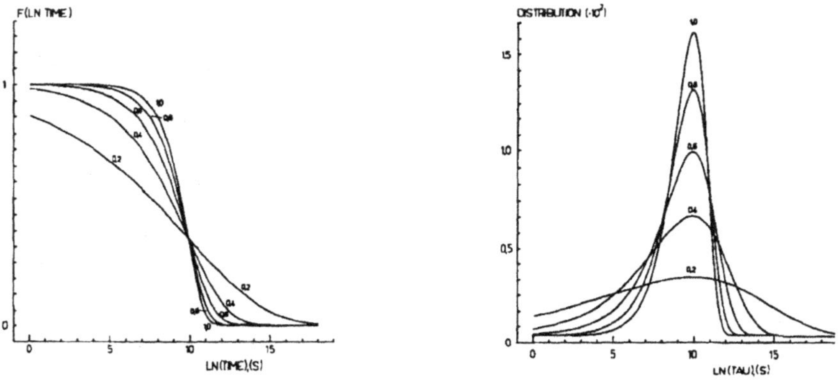

Fig. 4 The Kohlrausch response function $\phi(t) = \exp[-(t/\tau)^N]$ at $\ln\tau = 10$ and different exponents and the corresponding relaxation time distributions

with $\ln\tau = 10$. Such a function (referred to as Kohlraush-function) is found very often useful in describing mechanical as well as dielectric responses . The distribution function calculated on the basis of Equ. (25) is also shown in Fig. 4. The τ-value corresponding to the maximum of this curve may be considered as the "mean relaxation time" while exponent N determines the shape of the distribution. τ is usually strongly temperature dependent. By increasing temperature the distribution moves to shorter times.

It is found experimentally, [2,4], that the temperature dependence of τ for side-chain rotations and for local main-chain motions is Arrhenius-like; i.e.

$$\tau(T) = \tau_0 \exp(E_A/RT) \qquad (27)$$

where E_A is the activation energy of the transition, R is the gas constant, T is the absolute temperature. The shift of the relaxation time distribution with respect to a reference temperature T_r is described by the time-temperature shift factor

$$a_T = \tau(T) / \tau(T_r) = \exp(E_A/R)[(1/T) - (1/T_r)] \qquad (28)$$

In the glass-transition temperature range the Arrhenius equation does not work; instead the Williams-Landel Ferry (WLF) [6] equation becomes useful .

The corresponding shift factor is

$$a_T = \exp [A(T-T_r) / (B+T-T_r)] \tag{29}$$

where T_r is an arbitrary reference temperature above T_g, A and B are the WLF constants.
By increasing the temperature parameter τ is decreased (E_A and A are negative); the relaxation time distribution is shifted to shorter times.

PHYSICAL AGEING

From the viewpoint of radiation chemistry it is important to realize that solid polymers exhibit considerable molecular mobility even at low temperatures. In glassy polymers there is mobility not only because of local chain- or side-group movements but by the fact that a polymeric glass is practically never in thermodynamical equilibrium. The process of approaching equilibrium, evidently, involves rearrangement of structure. When a polymer is cooled down by about 20 °C below T_g after processing this structural rearrangement may proceed for years; so thermodynamic equilibrium cannot be reached in a reasonable period of time. As a result of this in a glassy polymer slow, but very significant, changes of the properties occur: the specific volume decreases (volume relaxation [7]), the enthalpy also decreases (enthalpy relaxation [8]) and the mechanical and dielectric relaxation time distributions are shifted to longer times i.e. opposite way as by increasing the temperature . The tendency of a glass to reach equilibrium is referred to as "physical ageing" [9] because it produces similar changes in the mechanical properties as chemical ageing without any change of the chemical structure.

Physical ageing is a general property of glasses no matter if they are organic or inorganic; polymeric or non-polymeric; it is observed in semicrystalline materials too, even above their amorphous glass transition temperature [10] .

The shift of the mechanical or dielectric relexation time distribution due to physical ageing can be measured and calculated as

$$\ln [a_v (t_s , t_{sr})] = C [\ln t_s - \ln t_{sr}] \tag{30}$$

where a_v is the shift factor of physical ageing, t_s is the storage time at constant temperature after processing, t_{sr} is a reference storage time, C is an experimentally determined parameter. Above T_g C=0 because there equilibrium is reached in a short time. Below T_g C is usually in-between 0.5 and 1.2 in a rather wide temperature range. This means that in this range one decade storage time would shift the relaxation distribution by about one decade to longer times. This is a considerable effect. As an illustration Fig. 5. shows the creep compliance of polystyrene at different storage times below T_g at constant temperature. It is seen that the compliance curves are shifted to longer times i.e. the mobility of the main chain segments decrease by storage . Evidently this should influence any kind of chemical reactions in the glass including radiation chemical ones.

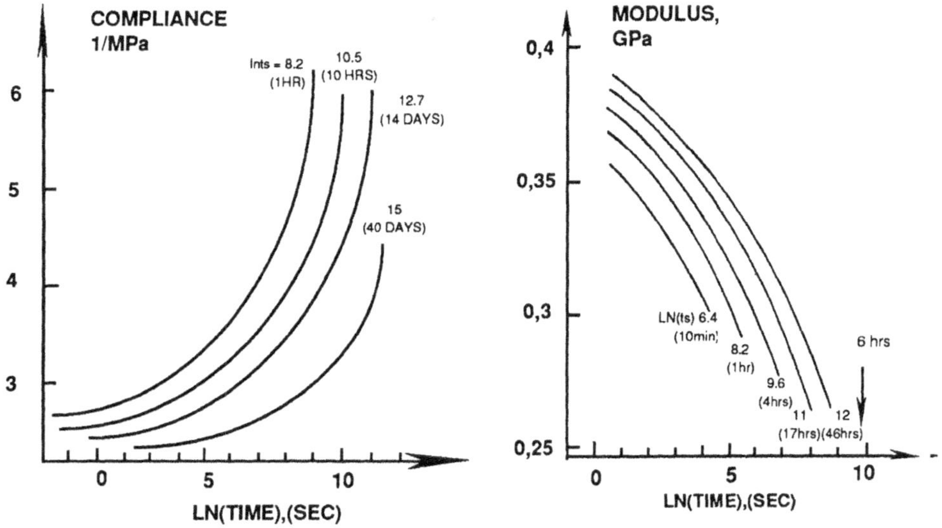

Fig. 5 a) Creep compliance curves of polystyrene measured at 40 °C after different storage times. b) Stress relaxation of polypropylene after annealing at 100 °C for 17 hours cooled to 40 °C and measured there after different storage times.

Fig. 5 also shows stress relaxation curves for semicrystalline polypropylene measured at 40 °C after annealing at 100 °C. 40° C is higher than T_g (-5 °C) but lower than the highest transition temperature of polypropylene below crystalline melting. The similarity of the storage-time effect with that of polystyrene below T_g indicates that in semicrystalline polymers there is physical ageing even above T_g [10].

Another important feature of physical ageing is that by increasing the temperature a part or all of the ageing effect may be erased even below T_g. The time needed for this erasure, when a glassy polymer after storage for a period of time t_{so} at temperature T_o is heated-up to temperature $T > T_o$, is given by Struik [9] as

$$t_m = kt_{so} \exp[\ H/R\ (1/T - (1/T_o))\ ] \qquad (31)$$

where k and H are material's constants, R is the gas-constant. For polystyrene e.g. k=0.18, H/R=14.2 . According to Equ. (31) t_m decreases by increasing T. Above T_g the previous history is at once completely erased.

The instantaneous state of physical ageing of a polymer can be characterized by the excess specific volume $\delta = [\ v(t) - v(>\infty)\ ] \ / \ v(\infty)$ which is time- and temperature-dependent; $v(\infty)$ is the equilibrium value. During the course of physical ageing the excess specific volume decays slowly towards zero. Change in the excess specific volume involves shift of the relaxation time distribution. Thus , the specific volume may be used as a measure of the large-scale mobility in the glass. By increasing the temperature this process is reversed: the specific volume i.e. the mobility increases to reach maximum then decreases again . The time needed for reaching this maximum is t_m in Equ. (31). This phenomenon is referred to as "memory effect". The excess specific volume (δ) is found experimentally a fair parameter for characterizing large-scale mobility involving structural changes. No matter what way one reaches a certain value of d the large scale mobility and the corresponding properties will be the same. The rule, however, does not hold for smaller-scale mobility, i.e. that of short chain segments and side-groups .

The "memory-effect" is illustrated in Fig. 6 for polystyrene stored at different temperatures (T_o) and then heated-up suddenly by temperature-steps indicated up to 96 °C where the volume relaxation experiments were performed. It is seen that if the temperature-step $\Delta T = T - T_o$ is large, the history of the sample is very soon erased and the volume, i.e. mobility, decreases with time. If, however, the temperature-step is small, the mobility increases with time to reach maximum. This range of increasing mobility may be totally inside our experimental time-interval which is usually in the range of 5-6 hours. In this interval the relaxation time distribution moves toward shorter times continuously during any kind of isothermal experiment provided that the storage time of the sample below the temperature of the measurement was long enough. On the other hand, if the storage-time is short, the maximum time t_m is reached in a short time; afterwards the volume continously decreases and, correspondingl, the relaxation time distribution is continuously shifted to longer times . This is referred to as simultaneous physical ageing because it proceeds along with the actual (real) time of the measurement. Evidently, even with long storage times, if the time of the measurement is sufficiently long, after a certain time there will be simultaneous physical ageing.

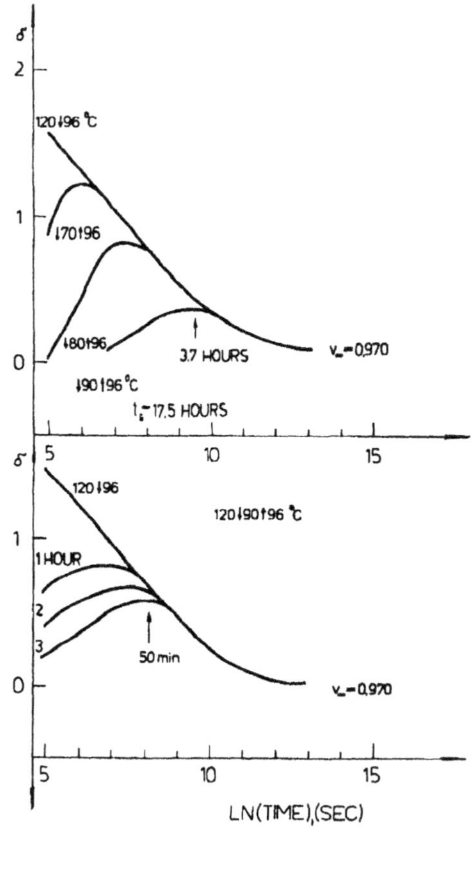

Fig. 6 Volume relaxation in polystyrene at 96 OC after storage for 17.5 hours at different temperatures (upper curves) and that after storage at 90 OC for different times (lower curves).

Creep and stress relaxation under the condition of simultaneous physical ageing can be described by replacing time t in the response function (Equ. 26) by the effective (reduced) time [9]

$$\lambda = \int_{t'=0}^{t} [t_s/(t'+t_s)]^c \, dt' = t_s/(1-C) \left[\left[1+(t/t_s)\right]^{(1-c)} -1\right] \qquad (32)$$

As a result creep or stress relaxation is slowed down; the corresponding apparent distribution is broadened in the long-time side. As an example Fig.7 shows stress relaxation curves for unplasticized poly vinyl chloride measured after a storage time of 17 hrs. These curves were constructed by using experimentally determined parameters and the Kohlrausch response function with the effective time according to Equ. (32). The dotted line corresponds to the case when simultaneous physical ageing was not taken into account. The distribution curve calculated according to Equ. (25) is also showh.

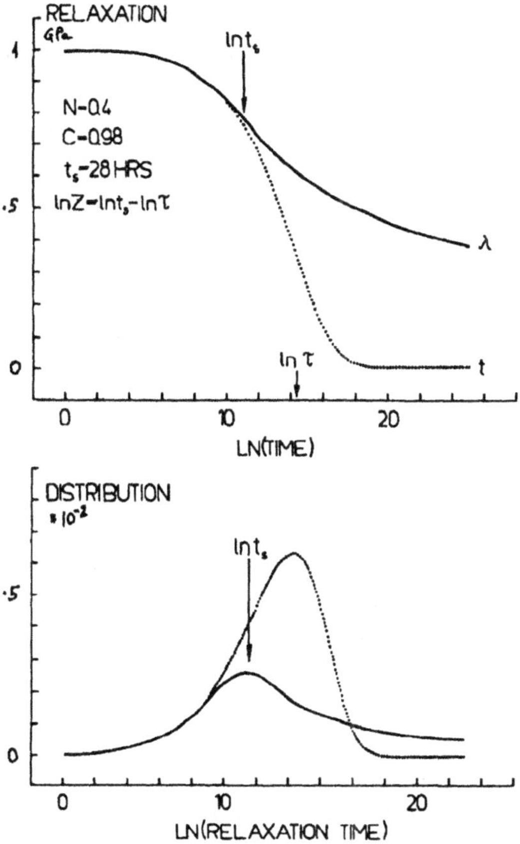

Fig. 7 Stress relaxation of unplasticized polyvinylchloride at 40 °C after storage for 17 hours. The curves were constructed by fitting the data to the Kohlrausch function considering real time t and volume relaxation effective time λ .

The broadening of the apparent distribution to almost infinite times indicates that there is a finite probability of large scale motion within any reasonable time-interval of the measurement.
To radiation chemist this means that in the glassy amorphous state there is a finite probability for reacting species to travel long-range. Reactions in the glassy or semi-crystalline polymers may be understood by this mechanism without the need of introducing quantum tunnelling or migration mechanisms.

EXPERIMENTAL TECHNIQUE AND EXAMPLES

As was discussed in the previous chapter in order to characterize the mechanical and electric properties of polymers most important is to know their response function to step-loading. At first this seems very simple. The problem is that usually one has to deal with several transitions each of them having a rather broad relaxation time distribution spectrum. In order to detect them very long measuring times would be needed in the time domain. As was also discussed previously, by Fourier transformation it is possible to transform the time-scale to the frequency-scale which is usual in spectroscopy. In principle, the frequency can be varied in a wide range, so the distribution in the frequency domain may be measured in a short period of time. The frequency domain technique is traditionally used in dielectric spectroscopy because the sinusoidal electric field "load" can be relatively easily changed by several decades. In mechanical spectroscopy the frequency domain method is technically difficult and is, correspondingly, rarely used.
Since the parameters of the response function are temperature dependent the relaxation time distribution can be continuously shifted to shorter times by increasing the temperature. This makes it possible to measure in the temperature domain i.e. at constant frequency or load-time by sweeping the temperature. This method is commonly used in mechanical spectroscopy and is also often used in dielectric spectroscopy. A very simple variation of the temperature domain method is to apply a constant load at low temperatures and, by sweeping the temperature upward, detect the reponse . Such a method is often referred to as "thermally stimulated" one.

Time-domain techniques
The time-domain method may be used in the mechanical as well as electric case. By application of a step-function load the response (mechanical stress

or strain or electric current) is measured. The time-interval covered depends on the fastness of the detecting system and on the total time of the measurement. This is the most straightforward way of measuring the response function. In principle this method can be used from a few nanoseconds up to years but in practice it is mainly used to detect slow mechanical or dielectric relaxation processes . A simplified scheme is shown in Fig. 8.

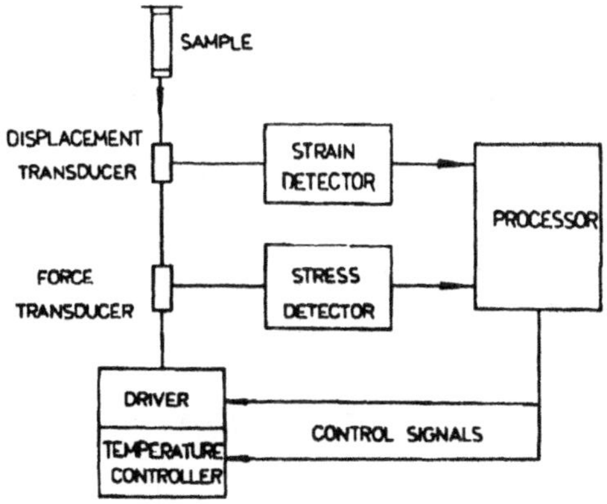

Fig. 8 Simplified scheme of a time domain mechanical spectrometer.

Since the time-scale of the measurement is technically limited, the best way is to fit an analytical response function, as e.g. the Kohlrausch function Equ . (26), to the measured data. For example in the case of mechanical stress relaxation the measured time dependent modulus , according to Equs. (5), (8) and (26), has four parameters: $E(o)$, $E(\infty)$, t and N. By using approximate values for these parameters by a gradient-fitting procedure [11] it is possible to find those values which would fit best to the experimental datapoints . This way an analytical form of the response is obtained which is easy to transform into the frequency scale, if needed.

An example is shown in Fig. 9. The experimentally measured tensile modulus-values of a plasticized poly(vinylchloride) compound (PVC) were measured as a function of ln (time). From the data by the gradient-fitting procedure the parameters of the Kohlrausch response function were

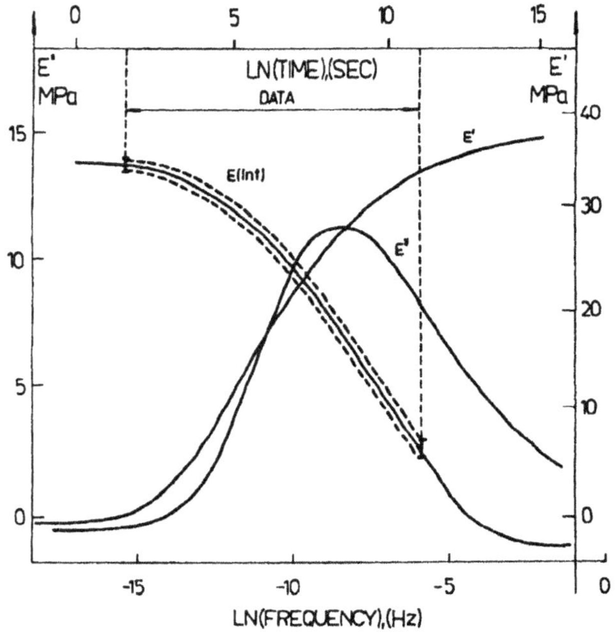

Fig. 9 Transformation of a stress relaxation curve into frequency dependent storage E' and loss E" moduli. Sample: plasticized polyvinylchloride, temperature: 40 °C. The experimental datapoints lie in-between the dashed lines.

obtained. The full line represents the corresponding curve. From this curve by numerical fast Fourier transformation [12] the storage E' and the loss E" moduli were calculated and plotted against the logarithm of frequency. The relaxation time distribution spectrum was also calculated by using Equ. (25). The detected transition is the glass-transition of plasticized PVC .

This method can be refined by repeating the measurement at different temperatures above dilatometric T_g. This results in a series of isotherms i.e. a series of parameters. Parameter τ of the Kohlrausch function is found to depend on the temperature; above T_g according to the WLF-equation (Equ. 29). From this experimentally determined dependence the WLF constants are determined. Knowing them one is able to calculate any stress relaxation E (t) or creep compiance D (t) function from dilatometric T_g (-20 °C in this case) up to about +150 °C , where another transition occurs.

In this particular case the temperature dependence of exponent N, which determines the shape of the relaxation time distribution function, is found not significant, i.e. the spectrum is just shifted by changing the temperature without changing shape. In such cases from isotherms by "horizontal" shifting a "master curve" can be constructed. By application of the gradient fitting procedure no graphical shifting is needed; the master curve can be constructed by determining parameters τ , N, E (o) and E (∞) .

As an example Fig. 10 shows stress relaxation curves of radiation crosslinked linear low density polyethylene (X-LLDPE) above the crystalline melting temperature. In this temperature range uncrosslinked polyethylene flows. Depending on the crosslink-density X-LLDPE above crystalline melting temperature is elastic; the deformation is almost completely recovered . From the stress relaxation curve the relaxation time distribution calculated by Equ. 25 is shown in Fig. 10 at different absorbed doses, i.e. crosslink-densities. The fact that there is a relatively long stress relaxation indicates the existence and importance of the entanglement network besides chemical crosslinks. The stress relaxation curves are well described by the Kohlrausch function.

Recovery of the deformation may be measured at different stages of the stress relaxation process by releasing the strain at different periods of times after loading and monitoring the time dependent recovery.

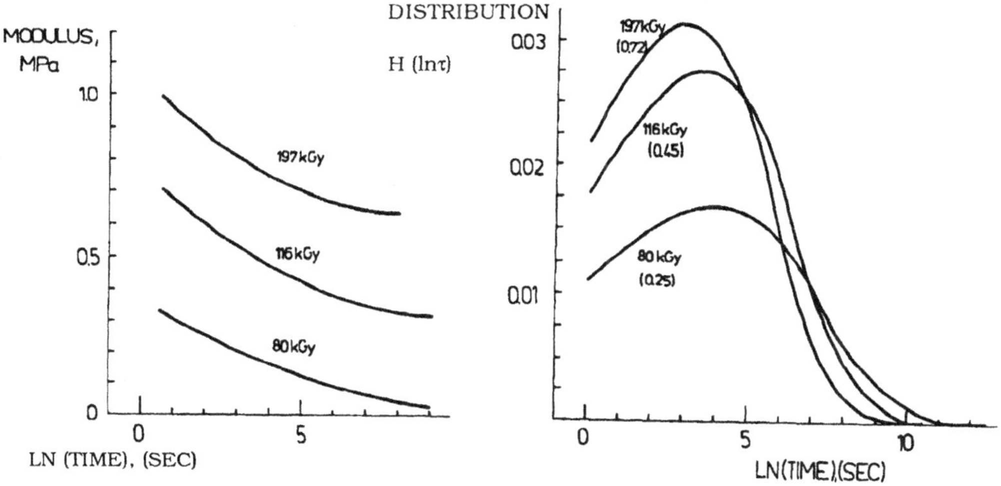

Fig. 10 The effect of irradiation on the stress relaxation of a linear low density polyetihylene (Mitsui ULTEZEX 2021 L) at 140 °C

This way parts of the distribution can be sampled. A series of such curves are shown in Fig. 11 for X-LLDPE. The recovery function is normalized to the original deformation and the irreversible part (flow) is subtracted. When the load time is short, just the initial part of the relaxation spectrum is sampled; the corresponding recovery relaxation time distribution, R (lnτ), is narrow. By increasing the loading time before recovery more and more of the relaxation time spectrum can be sampled.

The electric time-domain experiment can be performed similarly: by applying an external electric field the current is measured as a function of (time)-or rather log (time)-by using a scheme shown in Fig. 12. A fitting procedure, similar to that outlined before, may be used to determine the parameters. The current response to the electric field is a decaying function ("charge" current) which tends to a constant value I (∞) which is a result of "ohmic" conductivity. When the field is made off, a negative current ("discharge"- current) is observed which decays to zero.

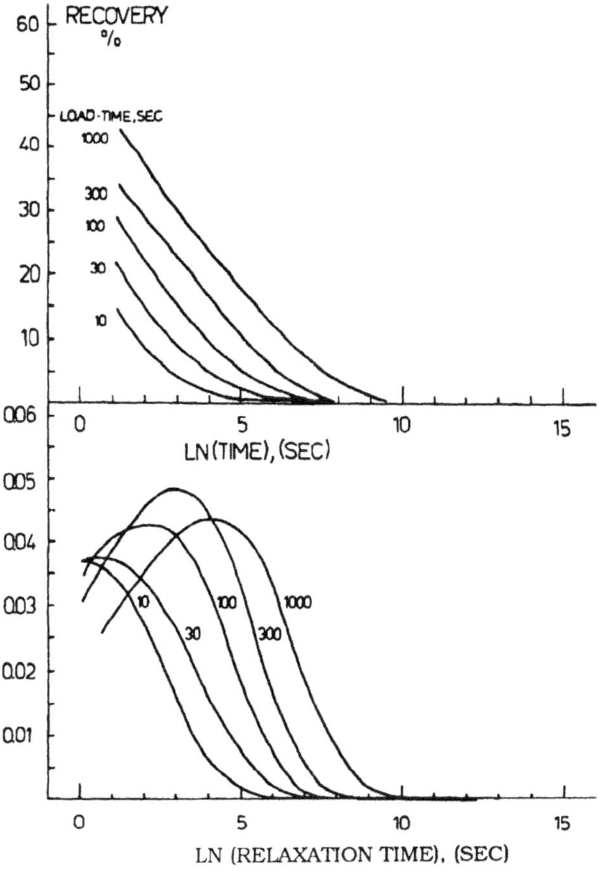

Fig. 11 Recovery of the deformation after different phases of the stress relaxation. Sample: Mitsui ULTZEX 2021 L type linear low density polyethylene, temperature: 140 °C, deformation: 4 %, irradiation dose: 336 kGy.

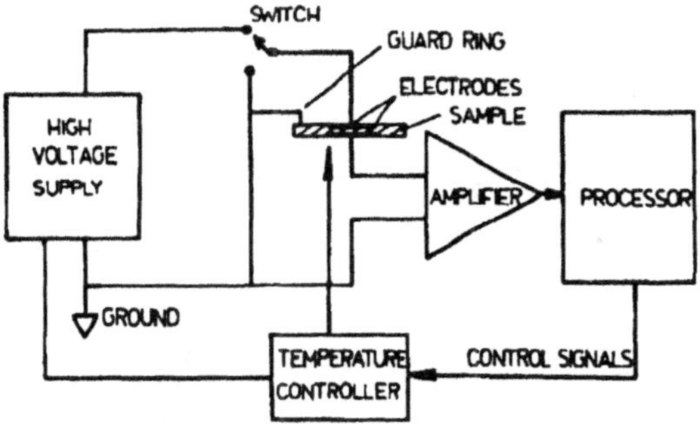

Fig. 12 Simplified scheme of a time domain dielectric spectrometer.

The charge period is analogous to creep while the discharge one to recovery keeping in mind that the response function is the derivative of the mechanical one.

The derivative of the Kohlrausch creep function is

$$d\left[1-\phi_K(t)\right]/dt = (N/\tau)(t/\tau)^{N-1} \exp\left[-(t/\tau)^N\right] \tag{33}$$

when $t \ll \tau$ the first term dominates in the current decay ($N<1$). Indeed the decay is experimentally often found a simple power function: $j(t) - j(\infty) \propto t^{-k}$ with $k = 1-N$, i.e. in the logarithmic scale the decay is approximately a straight line, as shown in Fig. 13. In order to transform time domain dielectric response to the frequency-domain a simplified numerical transformation may be used: [13]

$$\varepsilon'(\omega) - \varepsilon(\infty) = (t/C_oV_o)\left[(1000/t)\int_{t'=0}^{\sqrt{2}} I(t')dt' - 261.2\, I(t) - 1231\, I(4t)\right]$$
$$\varepsilon''(\omega) = (t/C_oV_o)\left[314.3\, I(t/2) + 2382\, I(2t) - 1301\, I(4t)\right] \tag{34}$$

where C_o is the geometrical capacitance, pF, V_o is the applied voltage (volts), I is the current in nanoamperes , $\varepsilon'(0)$ and $\varepsilon'(\infty)$ are the $\omega \to o$ resp. $\omega \to \infty$ limiting permittivities, $\omega = 1/t$.

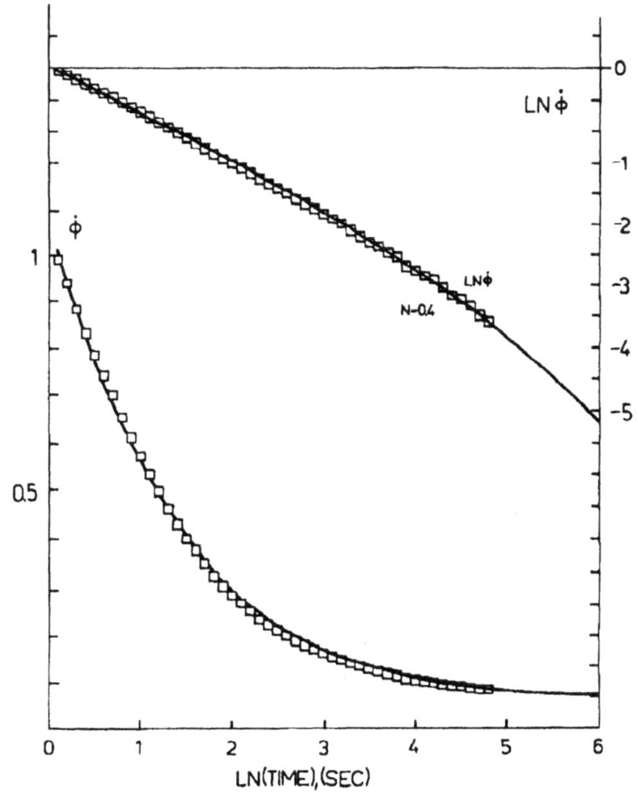

Fig. 13 Fitting the derivative of the Kohlrausch function to an experimental electric current decay curve. Sample: gamma irradiated low density polyethylene, temperature: 40 °C.

Fig. 14 e.g. shows a series of dielectric loss $\varepsilon''(\log \omega)$ curves obtained from charge-discharge experiments by transformation (34) for low density polyethylene gamma-irradiated in air with increasing doses. The increase of the loss is a result of oxidation.

The time domain technique in dielectric spectroscopy is also used in the very high frequency (GHz)- range but this is of little importance for polymers. It seems to be possible to cover the frequency range from 10^{-5} Hz up to about 10^{7} Hz in one run by the time domain method because the necessary fast-response electronic parts are available. In practice, however, the frequency band between 10^{-4} and 10^{4} Hz may be covered. The higher frequencies are usually measured by conventional "ac" dielectric spectroscopy using sinusoidal field. The reason of this is technical: at low

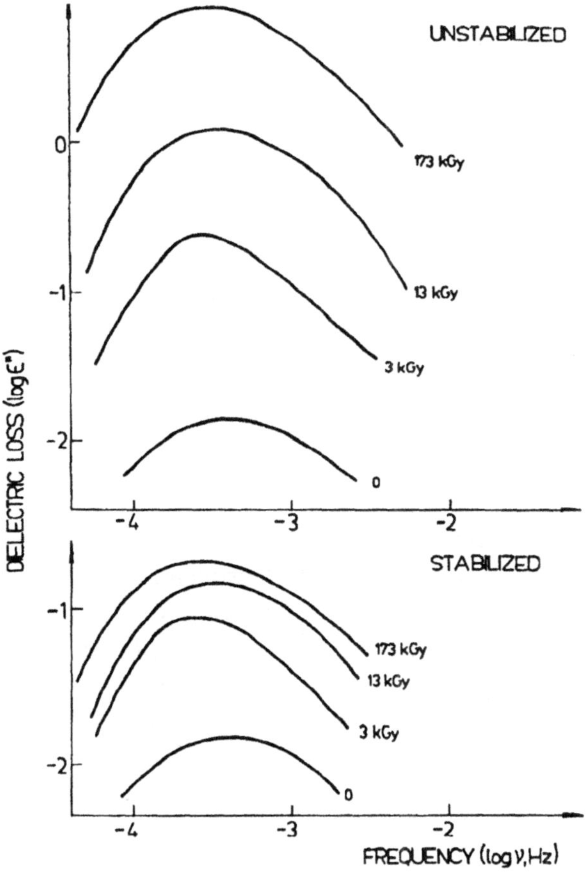

Fig. 14 Dielectric loss curves obtained from current decay for low density polyethylene irradiated in air by ^{60}Co gamma rays. The stabilized sample contained 0.1 % santonox anti - oxidant. Temperature: 40 °C.

frequencies relatively high time-constants are needed for the amplifiers to reduce noise level. At higher frequencies (short-times) the time constants should be reduced and no simple automatic way to do this has been developed yet .

Frequency-domain methods

Frequency-domain techniques involve measurement of the response of the material to sinusoidal (harmonic) loads as a function of the frequency. The response is also sinusoidal which can be divided into two components : one being in-phase with the load (real component) and the other is by 90° out-of-phase (imaginary component). Separation of these components may be done analog way by using a phase sensitive detector. In this case, correspondingly, instead of the time dependent response function $\phi_d(t)$ the Fourier transforms $f'(\omega)$ and $f''(\omega)$ are measured (cf Equ. 13). From this, by inverse Fourier transformation, the response function may be obtained.

In the mechanical case such a measurement is rather difficult because several decades of angular frequency should be covered in order to detect a single spectrum-band. In a frequency-domain mechanical spectrometer a mechanical, magnetic, or piezo-electric loader is used to apply a sinusoidal strain in extension. The strain, the force and the angular frequency are converted into electric signals by transducers. These signals are detected by a phase-sensitive detector which separates the force components "in-phase" and by 90° out-of-phase with respect to the deformation. The signals may be printed out or just shown on a digital voltmeter display. There are many variations of this method but the principle is the same. For details see Ref [2].
In order to do a frequency domain experiment the frequency of excitation should be changed by several decades, which is technically difficult. In the dielectric case the experimental arrangement is similar to that of Fig. 12 with the difference that the sinusoidal current-response signal is phase-detected by using the voltage as reference. Since the current response-function is the derivative of the polarization one, the real part of the current is proportional to the imaginary part of the permittivity (dielectric loss) and the imaginary part of the current is proportional to the real part of the permittivity.

Sinusoidal dielectric excitation is technically easy to realize from about 0.1 Hz up to 10^5 Hz. At high frequency range up to about 10^{10} Hz it is also possible to do dielectric experiments but the technique is not at all easy . Nevertheless, with some technical difficulties, frequency-domain dielectric spectroscopy is capable to cover the frequency range from about 0.1 Hz up to 10^{10} Hz, i.e. 11 orders of magnitude. By combination with the time domain technique and with the far infrared Fourier transform (interferometric) infrared spectroscopy it is principally possible to cover the frequency range from 10^{-5} Hz up to about 10^{14} Hz.

Temperature-domain methods

As was mentioned earlier the relaxation time distribution spectrum is shifted to shorter times by increasing the temperature. Thus, by keeping the

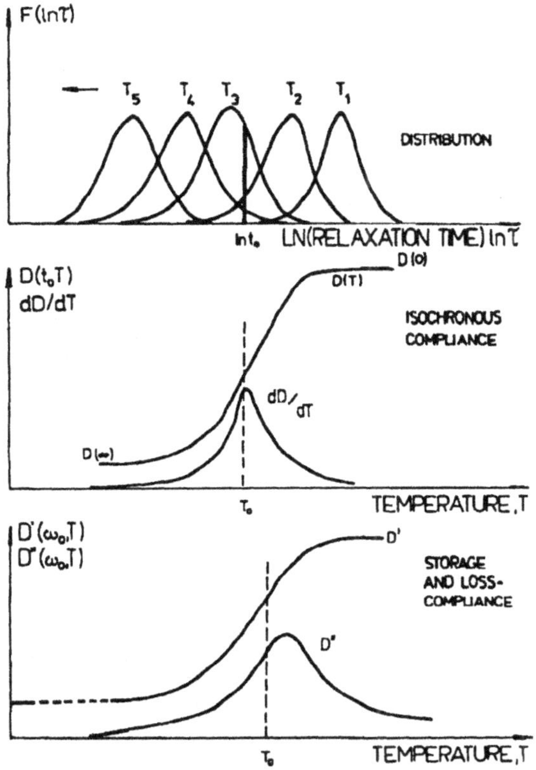

Fig. 15 Illustration of a temperature domain mechanical spectrum-band.

creep or stress relaxation time (or the angular frequency) constant , "isochronous" compliances or moduli may be measured by sweeping the temperature or by changing it stepwise. This situation is schematically illustrated in Fig. 15. By shifting the temperature the distribution travels through the "experimental window" represented by the load-time t_0. As a result the compliance (or modulus) exhibits a dispersion-like curve the low temperature side of which corresponding to short relaxation times while the high temperature side to longer ones.

The peak of the relaxation time distribution represented as a function of the temperature depends on the load-time or -frequency. This method is very often used in "dynamic mechanical analysis", DMA [14] , applying sinusoidal excitation at constant angular frequency. The resulting storage-and loss-compliances are plotted against temperature which is swept upward. By increasing the frequency the loss peak is shifted to higher temperatures.

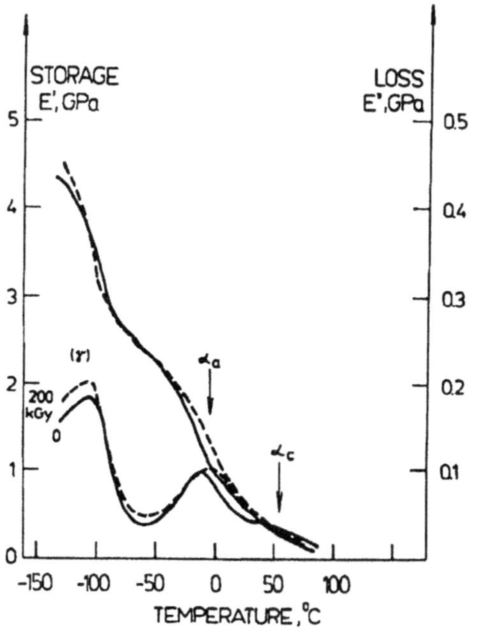

Fig. 16 Temperature domain (dynamic mechanical) spectrum of a linear low density polyethylene (Mitsui ULTZEX 2021 L) showing the effect of irradiation in air. Frequency: 10 Hz, rate of heating: 5 °C / min.

As an illustrative example Fig.16 shows a typical DMA spectrum of low density polyethylene measured at a frequency of 10 Hz (ω = 628) as a function of the temperature. In this case stress relaxation-mode was used, so the tensile storage (E') and loss (E") moduli are shown. The spectrum exhibits three main transitions labelled by α_c ("crystalline"), α_a ("amorphous") and γ (local motion). For details see Refs [2] , and [4].

By irradiation in the absence of oxygen PE becomes crosslinked. The effect of crosslinking to the mechanical spectra is, however, very small in comparison to that of the change of the crystalline structure, " morphology" . Thus, the changes of the DMA spectra observed after irradiation are mainly due to the changes of morphology caused by the chemical processes (crosslinking or degradation). When e.g. the irradiated polymer is annealed above its crystalline melting temperature, by recrystallization the morphology becomes drastically different from the unirradiated material heat-treated the same way. By irradiation the spectrum is only slightly affected. After recrystallization the α_a and α_c transitions are significantly changed with respect to those of the unirradiated sample crystallized the same way. Similarly by applying sinusoidal electric excitation of constant angular frequency the real and imaginary parts of the dielectric permittivity

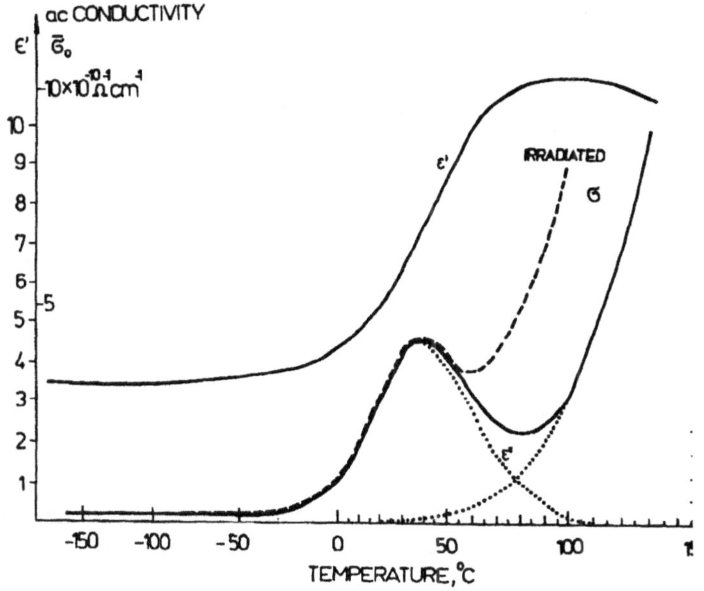

Fig. 17 Temperature domain dielectric spectrum of plasticized polyvinylchloride showing the way of separating dielecktric loss from conductivity. Frequency: 1 kHz, rate of heating: 2 °C / min, irradiation: 20 kGy ^{60}Co

may be measured as a function of temperature resulting in similar curves as shown in Fig.16. Fig.17 shows a typical temperature domain dielectric spectrum of plasticized polyvinylchloride /PVC/ at 1 k Hz. The real part of the dielectric permittivity and the total ac conductivity is plotted against temperature. The permittivity versus temperature curve exhibits a transition at the glass-rubber transition temperature of the compound. This is very slightly affected by irradiation. The total ac conductivity curve is a superposition of the exponentially increasing "ohmic" part, which is not frequency-dependent, and the dielectric absorption i.e. the temperature- and frequency-dependent imaginary part of the permittivity.

By irradiation the ohmic part increases dramatically as a result of radiation induced degradation, while the dielectric loss part is hardly affected. In order to interpret temperature domain curves, for simplicity, let us assume that in the Kohlrausch function Equ. (26) N=1 , τ (T) is Arrhenius-like and , further, that the limiting moduli are independent on temperature. This is, of course, over-simplification but makes it easier to demonstrate the main tendencies. The Fourier transforms of Equ. (26) in this particular case can be expressed in analytical form, so the temperature dependence of the storage-and loss-moduli are expressed as

$$E'(\omega_0 T) = E_\infty + (E_0 - E_\infty)\left[(\omega_0 \tau)/(1+\omega_0^2 \tau^2)\right]$$

$$E''(\omega_0, T) = (E_0 - E_\infty)\left[(\omega_0^2 \tau^2)/(1+\omega_0^2 \tau^2)\right] \tag{35}$$

The maximum of the E" (ω_0,T) function is when $\omega_0 \tau (T_0) = 1$; T_0 may be referred to as the transition temperature at ω_0 . By increasing ω_0 , $\tau(T_0)$ must decrease i.e. T_0 increases. By introducing the temperature dependence of τ (e.g. Equ. 27) the temperature dependent moduli and compliances as well as the dielectric permittivities can be analytically described.

In fact, of course, N ö1 and N, as well as the limiting moduli, are temperature dependent. Consequently temperature-domain mechanical or dielectric spectra are always difficult to evaluate quantitatively.

The microcomputer-aided fitting-procedure is very helpful in such cases too. As an illustrative example Fig. 18 shows a series of temperature-domain

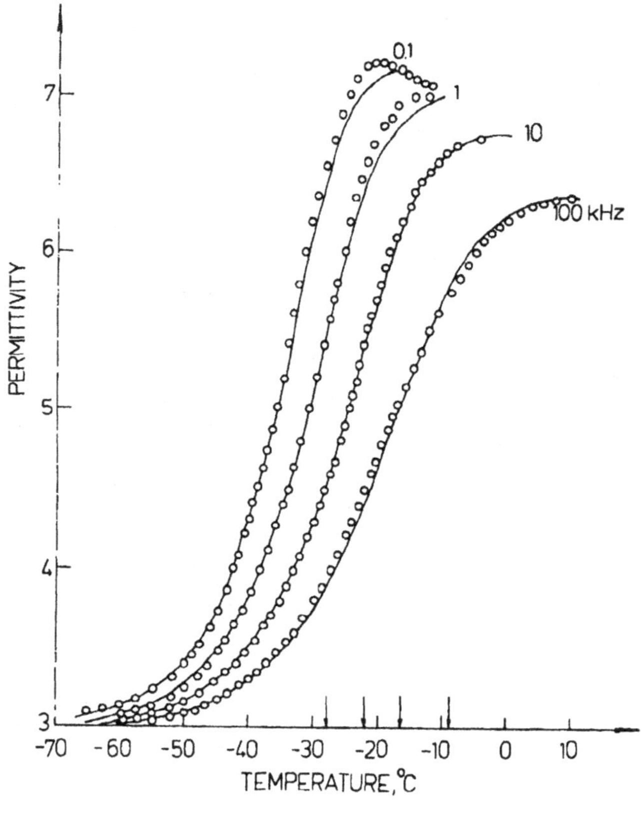

Fig. 18 Temperature domain dielectric spectra of a polyester of 1,6 hexane diol - maleic anhydride with 30 % styrene. The experimental curve was fitted to the Geny - Monnerie (17) autocorrelation function (full lines).

dielectric spectra of a mixture of an unsaturated polyester resin with styrene measured at different frequencies in one run [15]. The experimental spectra (circles) were fitted to a response function introduced by Gény and Monnerie [16]. It is seen that the fit is fairly goog: the temperature dependence of the limiting permittivities could also be taken into account. Such a fitting procedure was also found useful for describing dielectric transitions in radiation cured polyester resins [15].

By curing resins or by vulcanization of rubbers it is possible to keep the load-time (t_0) or the frequency (ω_0) and the temperature (T_0) constant and measure the change of the mechanical compliance or modulus or dielectric permittivity as function of the reaction time. In such experiments the relaxation time distribution is shifted by crosslinking to longer times making the compliance or permittivity decrease. The situation in this case is similar to that illustrated in Fig. 15 but the distribution moves opposite way (to longer times) as a result of crosslinking and its shape is drastically changed.

Fig. 19 shows a series of temperature-domain isochronous compliance curves measured at different stages of radiation crosslinking. It is seen that the transition is shifted to higher temperatures (longer relaxation times) and becomes broader by increasing crosslink density. Also the post-transition compliance-values are drastically reduced by crosslinking (modulus-values increased).

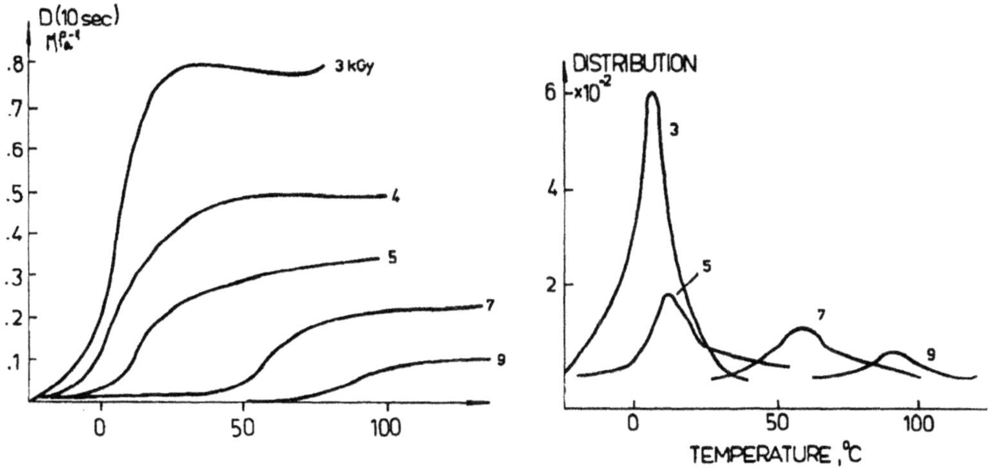

Fig. 19 Temperature domain isochronous creep compliance curves of an unsaturated polyester resin measured at different stages of radiation crosslinking. Load time: 10 sec, rate of heating: 1 °C / min, radiation: ^{60}Co gamma.

Temperature-domain measurements made by constant frequency (sinusoidal) or constant load-time (square-pulse) methods may be performed by stepwise or linear temperature-program; the rate of heating is only limited by the heat-capacity of the sample . By applying constant load (stress or strain) at low temperature and heating the sample up at constant rate (r) "thermally stimulated"creep or stress relaxation curves are obtained The electrical analogue (to creep) is the thermally stimulated polarization current curve measured at constant voltage sweeping the temperature upward linearly. Another possibility is to cool a material down in a strained state then, after removing the strain, measure the recovery (shrinkage) as a function of temperature without any load. The electrical analogue of this method is referred to as thermally stimulated dielectric depolarization (TSD) [13]. All these thermally stimulated methods are technically simple but quantitative interpretation of the obtained curves is rather complex . In the mechanical case a differential technique [18] is recommended in order to cancel-out the thermal dilatation of the sample and holder. One sample in this measurement is loaded (simply by weight or pneumatically) the other (reference) is unloaded. The true strain is calculated as

$$\varepsilon_t = \ln \left[1+(\Delta l_s/l_{so}) - (\Delta l_r/l_{ro})\right] \tag{36}$$

where Δl_s and Δl_r are the deformation of the sample and reference respectively measured as a function of temperature, l_{so} and l_{ro} are the corresponding lengths at the starting (lowest) temperature .

Such an arrangement, equipped with digital data-processing facility, is useful to perform any thermally stimulated or pulsed-load experiments with high accuracy. Thermally stimulated electric experiments may be performed by using the simple scheme shown in Fig. 12 equipped with a temperature controller and X-Y recorder.

Thermally stimulated mechanical or electric curves are interpreted by introducing an effective time

$$\lambda = 1/r \int_{t'=0}^{t} a^{-1}_T \quad (T'T_0) \tag{37}$$

instead of real time into the corresponding equations. In Equ. (37) r is the rate of heating (K/sec), T_0 is the starting temperature a_T is the time-temperature shift factor corresponding to the reference temperature T_0. Using this, thermally stimulated tensile creep is described as

$$D(\lambda) = D(o) + [\, D(\infty)\text{-}D(o)] \;\; \phi_\varepsilon(\lambda) \tag{38}$$

As the effective time l depends on the temperature according to Equ. (37) the temperature-dependent compliance can be calculated if a_T and ϕ_ε are known.

This simple method, however, works only when the limiting values D(o) and D(∞) are independent on the temperature and the conditions for shifting are fulfilled. In this case the temperature-scale can be converted into time-scale. The limitations are, however, very severe. Nevertheless the method is very useful for detecting transitions because its effective frequency is very low depending on the rate of heating mainly.

Fig. 20 shows the thermally stimulated dielectric depolarization (TDS) spectrum of radiation crosslinked low density polyethylene (X-LPDE). As was discussed previously the current is proportional to the derivative of the response function so the derivative of the creep curve is similar to the TSD-spectrum. Although polyethylene is nonpolar, all the transitions are detected. The current results from the dipoles formed by oxidation during the course of processing and irradiation.

The thermally stimulated technique offers a unique way for studying relaxation time distributions. The two main possibilities are shown in Fig . 21. The first one, referred to as "thermal sampling" of the distribution is performed simply by loading the sample at a temperature within the transition range and quench. This way a part of the spectrum (shaded area) will be frozen-in and released by the subsequent linear heating-cycle when the temperature derivative of the strain is measured without external load. By repeating the experiment at different straining temperatures more and more of the distribution can be sampled. The other method is similar but only a small part of the distribution is frozen-in: a "temperature window" . This way the whole distribution can be sampled in small parts. Both methods can be applied in the mechanical as well as in the electric case. For a recent discussion see Re. [18].

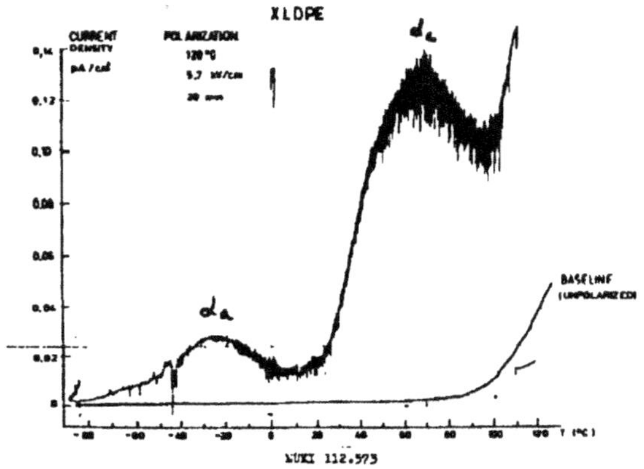

Fig. 20 Thermally stimulated dielectric depolarization current spectrum of low density polyethylene. Polarization: 5.7 kV/cm at 120 °C, rate of cooling: 3 °C/min, rate of heating: 1 °C/min in Argon atmosphere.

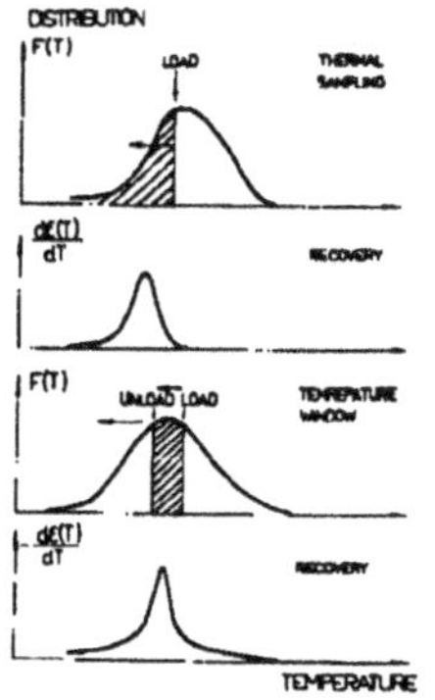

Fig. 21 Illustration of the method of thermal sampling of a dielectric or mechanical transition.

THERMALLY STIMULATED ELECTRICAL CURRENT-FLUCTUATIONS

By irradiation a lot of charge carriers become trapped inhomogeneously in the solid polymer. By thermal activation these carriers would produce thermally stimulated current (TSC) by heating the sample up without external field or in the presence of a small field. After this treatment the sample is "cleaned-up" from the trapped charges: no such current peaks are observed by subsequent heating-cooling cycles.

Analysis of the detrapping current-peak may be interesting in itself because of the possibility of determining the distribution of the trapping sites. Reversible current-fluctations coming from the sample can be observed in the TDS-spectrum of radiation crosslinked low density polyethylene (Fig. 20). This thermally stimulated depolarization current spectrum was obtained after previous electrical "cleaning" and after annealing the sample during polarization in order to stabilize a certain microcrystalline structure. This way the depolarization curve is completely reproducible; the transitions

marked by α_a and
α_c correspond to those measured by DMA (cf Fig. 16) taking into account the difference in the effective frequencies. The current-fluctations, which are superimposed to the depolarization-peaks in Fig. 20, are found also reproducible.

The fluctations can be observed without any previous polarization by increasing the temperature linearly or stepwise under shortcircuit conditions. An example for this is shown in Fig. 22. In this experiment the thermally treated, irradiated , and electrically cleaned sample was heated-up at a rate of 1 °/min without any external electric field and the shortcircuit current was measured. Fig. 22 shows on analog recording for illustration but in the actual experiment the current fluctations were fed into a microcomputer for analysis of their amplitude-distribution and frequency spectrum. The analog picture is, probably, more illustrative to show that fluctations start at about-10 °C where the α_a transition is located; its average intensity reaches maximum in the α_c-transition range and vanishes completely after passing through the crystalline melting temperature. By cooling the current-fluctations reappear at the temperature of crystallization.

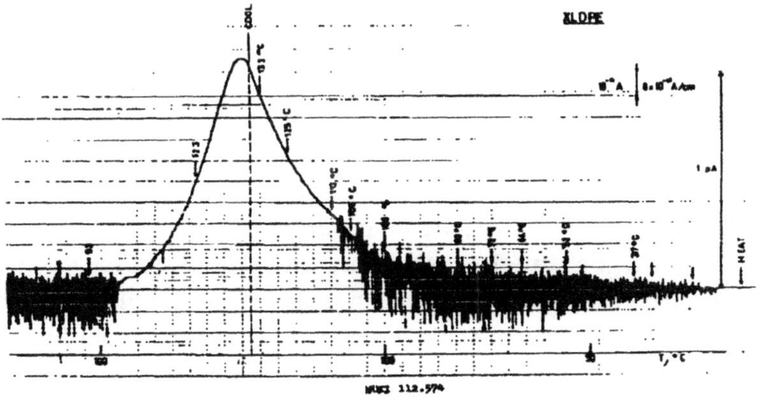

Fig. 22 Thermally stimulated current fluctuations in radiation crosslinked low density polyethylene measured after previous electrical cleeaning. Rate of heating : 1 °C/min Fluctuations vanish by passing through the melting temperature and re-appear by cooling.

The observed current-fluctations are interpreted as being due to large-scale motion of oxidized chains in the inter-lamellar space, governed by the rearrangement of the structure rather than local motion, because the most intense Fourier coefficients are at very low frequencie. Experiments made with stepwise temperature-jumps support this view: fluctations are drastically reduced after the structure reaches equilibrium after the jump [20].

CONCLUSIONS

The effect of high energy irradiation on the mechanical and electric properties of polymers is interesting from two basic viewpoints. One is radiation damage, i.e. deterioration of the properties by irradiation. This is important in such cases when plastic products are subjected to radiation, e.g. in radiation sterilization, in cases of equipment made of plastics working in the presence of radiation (reactors). The other aspect is radiation processing involving radiation polymerization, crosslinking (curing, vulcanization) and grafting.

By studying radiation damage mechanical properties are of basic importance because in every application there are requirements about mechanical properties. Radiation degradation in the presence of oxygen is known to exhibit dramatic effects to the mechanical properties. Poly (tetrafluoro-ethylene) (Teflon) e.g., which exhibits excellent thermal stability , decays into a powder by irradiation in air at a dose of a few kGrays. The reason of this apparent discrepancy is discussed in Rf. [21]. Other polymers as e.g. polyethylene, polystrene are rather resistant to irradiation.

In order to characterize radiation damage to the mechanical properties first the limits of linearity have to be determined. A practical rule is that the load in use should be much less than the yield-stress or-strain in the service-temperature range. Sometimes radiation damage is so high that there is no yielding; rupture takes place after a short quasi-linear $\sigma(\varepsilon)$-curve section. The Young modulus or stress relaxation modulus is found not too much affected by degradation. The average molecular weight must drop tremendously to see a significant change in the modulus (or compliance) or in the relaxation time distribution.

The electrical properties are rather sensitive to radiation. These are important in such cases when electrical insulators (cables, capacitors a.s.o.) are used in radiation field. Radiation damage involves not only reduction of the molecular weight by main chain scissions but conjugation and also by splitting of by-products which might influence the electrical properties (as HCl in PVC) appreciably.

In the case of radiation polymerization, grafting, crosslinking, curing , vulcanization the mechanical and electric tests are of vital importance. The crosslinking (curing) reaction e.g. can be monitored in situ by the dielectric technique which can be combined by differential thermal analysis [16].

By combination of the dielectric and mechanical techniques the products can be characterized even in the presence of reinforcing materials, fillers or dyes very efficiently.

REFERENCES

1. A. Charlesby: Atomic Radiation and Polymers, Pergamon Press; Oxford , London, New York 1960.
2. N. G. McCrum, B. E. Read , G. Williams: Anelastic and Dielectric Effects in Polymeric Solids. John Wiley & Sons, London, New York, Sydney 1967 .
3. I. L. Hopkins, R. W. Hamming; J. Applied Phys. 28, 906, /1957/
4. P. Hedvig, Dielectric Spectroscopy of Polymers, Adam Hilger Ltd, Bristol, Halsted Press, New York, /1977/
5. Yu. V. Zelenev, A. P. Molotkov : Vysokomol. Soed . 6, 8, 1426 /1964/
6. M. L. Williams, R. F. Landel, J. D. Ferry: J. Amer. Chem. Soc. 77, 370/1955/ See also in Refs 2 and 4
7. A. J. Kovacs: Makromol. Chem. Suppl. 81, /1979/
8. I. M. Hodge, A. R. Berens: Macromolecules 15, 762 /1982/
9. L. C. E. Struik: Physical Aging in Amorphous Polymers and Other Materials . Elsevier Amsterdam, Oxford, New York /1978/
10. C. K. Chai, N. G. McCrum: Polymer, 21, 706 /1980/, L.C.E. Struik, Polymer, 28, 1521/1987/
11. Ph. R. Bevington : Data Reduction and Error Analysis for the Physical Sciences, McGraw Hill, New York, San Francisco, St. Louis, Toronto, London, Sydney, /1969/
12. L.C.E. Struik, R. F. Schwarzl : Rheologica Acta 8, 134 /1969/
13. J. van Turnhout:Thermally Stimulated Discharge of Polymer Electrets, Elsevier, Amsterdam /1974 /

14. B. E. Read: Polyme. 22. 1580 /1981 /

15. T. Murayama: Dynamic Mechanical Analysis of Polymeric Materials. Elsevier. Amsterdam. Oxford. New York / 1978 /

16. Z. Jelcic. P. Hedvig. F. Ranogajec. I. Dvornik: Die Angewandte Makromolekulare Chemie. 21 . 2070 / 1985 /

17. F. Gény. L. Monnerie: J. Polym. Sci. Phys. Ed. 15. 1 /1977/

18. P. Hedvig: Differential Thermomechanical Analysis of Polymers in "Applied Polymer Analysis and Characterization" /ed. J. Mitchell. Jr./ Hanser Publishers. München /1987/

19. N. G. McCrum: Polymer. 25 . 299 /1984/

20. G. Bánhegyi. P. Hedvig in the "Proc. of the 6th Symposium on Radiation Chemistry" /eds P. Hedvig. L. Nyikos. R. Schiller /Academic Publisher. Budapest. 1987 . Vol. I.. p. 457.

21. P. Hedvig: J. Polymer Sci. Part A-1. 7. 1145 /1969/

THERMALLY STIMULATED DISCHARGE AND DIELECTRIC MEASUREMENTS OF CURE IN RESINS AND GELLED MATERIALS

RICHARD A. PETHRICK
Department of Pure and Applied Chemistry
University of Strathclyde
Cathedral Street,
Glasgow G1 1XL

INTRODUCTION

Network formation leads to a modification of both the short and long range relaxation properties of the matrix. Dipole relaxation will be dominated by the reorientational relaxation of polar elements of the polymer network structure. In a small molecule, dipole reorientation occurs at high frequency and is associated with the structural relaxation process of the fluid. In theory, the randomization of the dipole should occur via all angles, however, even in the case of small molecules, the interactions with neighbouring molecules restrict this process. In a low molecular weight polymeric species, the relaxation of the dipole can occur via a process which is analogous to that observed in small molecules and involves overall rotation of the whole molecule. As the molecular mass is increased to the local viscosity of the fluid rises and the net result is a slowing down of the rotational motion. The process observed in the low molecular species is similar to the first normal mode relaxation process, discussed in the chapter on ultrasonics and viscoelastics. The relaxation time is directly related to the end to end distance and hence to the root of the molecular mass. Increase of the molecular weight causes a decrease of the relaxation frequency. The dipole relaxation leads to a frequency dependence of the complex permittivity of the media * which can be written as

$$(\epsilon)^* = (\epsilon)' + i(\epsilon)'' \qquad (1)$$

where $(\epsilon\ )'$ and $(\epsilon\ )''$ are respectively the real and imaginary parts of the permittivity and are often referred to as the dielectric constant and loss. The width of the loss peak is a measure of the correlation of the relaxation of one dipole with another. A highly correlated process will exhibit a broad relaxation feature, whereas an weakly correlated dipolar relaxation process will approximate to the ideal Debye process. Alternatively, the occurance of several relaxation processes which have slightly different activation energies or pre-exponential factors can also lead to the observation of a broad relaxation process. The analysis of the curves in terms of distribution parameters has been made by Cole and Cole, Cole and Davidson and Havryliak and Negami.

O. Güven (ed.), Crosslinking and Scission in Polymers, 129–137.

The former of the three types of equation are presented below:

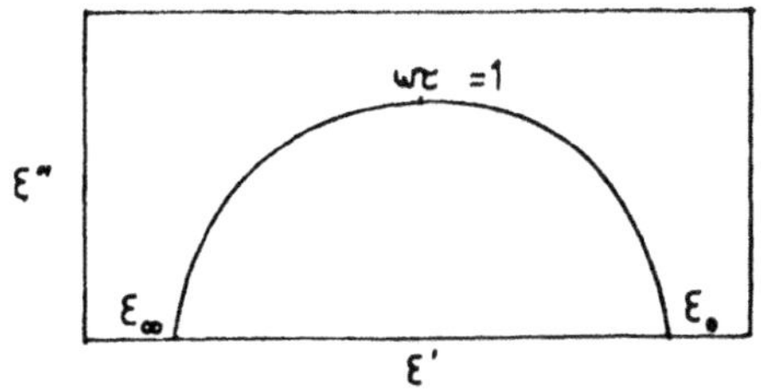

The Cole-Cole equation introduces a constant α.

$\epsilon^*(\omega) = \epsilon_\infty + (\Delta E)/(1+(i\omega\tau)^\alpha)$

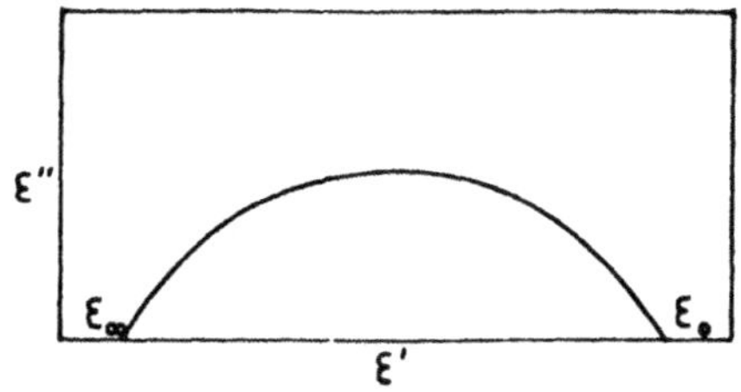

The Cole-Davidson equation introduces a constant β.

$\epsilon^*(\omega) = \epsilon_\infty + (\Delta E)/(1+i\omega\tau)^\beta$

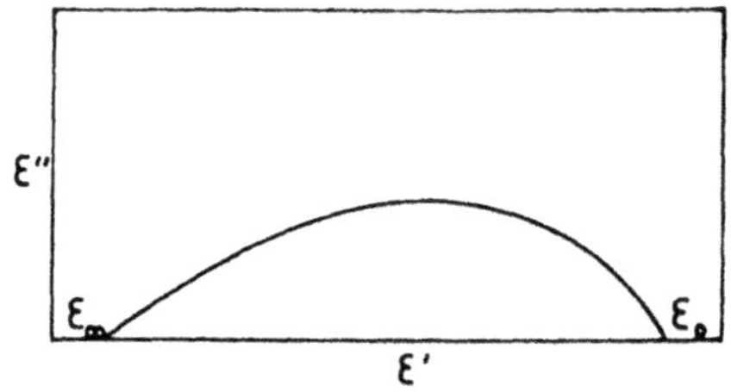

The Havriliak-Negami equation combines these two equations.

$\epsilon^*(\omega) = \epsilon_\infty + (\Delta E)/(1+(i\omega\tau)^\alpha)^\beta$

An alternative approach has been proposed by Williams based on the concept of correlation theory and this leads to the conclusion that the relaxation spectrum is a composite of self and auto correlation functions. The self correlation function is the mathematical representation of the memory of a dipole of its original position $\mu(i).\mu(i)$, whereas the auto correlation function is a correlation of the dipole with that of its neighbours.$\mu(i).\mu(j)$. It is not appropriate to develop the theory of dielectric relaxation in this article, however like ultrasonics and viscoelastics it is sensitive to both local and longer range dipolar motions.

Molecular Weight Effects If we compare the relaxation frequencies of narrow molecular weight samples of polyethyleneoxide, we observe for low molecular weight materials a very fast relaxation which is dependent on the end-to-end distance of the polymer coil, figure (2). As the molecular weight is increased the frequency decreases until above a certain molecular weight it becomes independent of molecular weight.

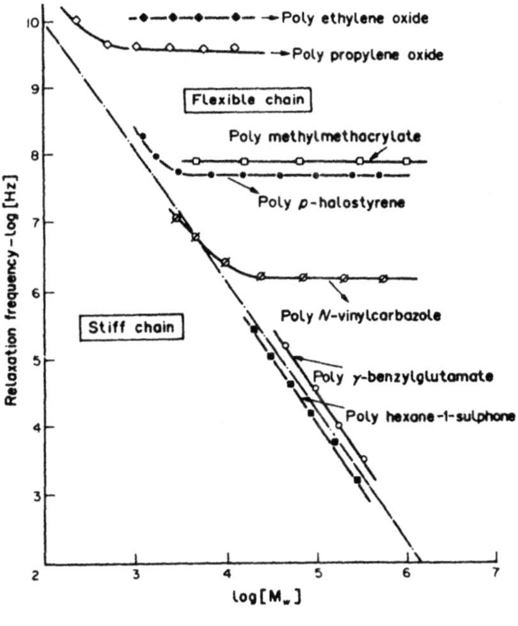

Figure(2) Variation of the dielectric relaxation frequency with molecular weight for polymers of different flexibility

A comparison of the data for the polymer indicates that in the molecular weight dependent region the dominant process is overall rotation of the polymer whereas in the molecular weight independent region it is reorientation of small segments of the backbone which leads to dipolar relaxation. Matsoui and Stockmayer have observed that two processes can be observed, one due to the local motion of the dipole-segmental reorientation and the other due to co-operative motion of large elements of the chain and having a frequency which is close to that for normal mode relaxation. Whereas in the case of viscoelastic or mechanical relaxation, the co-operative motion is dominant, in the case of dielectric relaxation it is local motion of the chain which dominates. Figure (2) does however illustrate the point that the relative importance of each of these processes depends on the stiffness of the polymer backbone. In a chain, such as polyethylene oxide, the molecular weight independent region starts at approximately 10^3, whereas in the case of polystyrene it is 10^4 and in the case of polyvinyl carbazole 10^5-10^6.

INFLUENCE OF GELATION OF POLAR MEDIA

To illustrate the way in which dipolar relaxation is influenced by the process of gelation, we will initially consider the case of a crosslinked polyester coating. Observation of the dielectric relaxation at a frequencey of 1KHz as a function of cure time provides information on the way in which the local mobility changes with network formation. Initially the dipolar relaxation occurs at high frequency and hence the loss is quite low. As the cure proceeds the molecular weight of the chains will increase and this will lead to a slowing down of the dipolar relaxation process, observed as an increase in the loss. However, as the molecular weight increases further we know that the longer range motions will be inhibited and the dipolar process will be dominated by local relaxation processes. As the network formation increases so the chain-chain interactions increase and the mobility of the dipole decreases. The result of this and also the reduction in free volume associated with the glass transition temperature leads to a reduction in the dipolar loss. The observation of the dielectric loss at 1 kHz is therefore an indication of the extent to which the chains in the system are being constrained by the formation of the matrix structure.

The thermally stimulated current method is also a useful method for the examination of the properties of a curing system. It is observed that in a simple epoxy resin with the exact stoichiometric ratio the TSD curves are composed of a single glass transition peak and a higher temperature space charge peak. The latter is associated with the motion of charge carriers through the matrix and are not associated with individual dipoles. However slight changes in the cure conditions and also the introduction of filler particles into the matrix lead to a multiple glass transition peak being observed, figure (3). The peaks correspond to regions of greater or less mobility. The observation of multiple glass transition peaks can be associated with regions which have a greater or lesser degree of crosslinking than the average value for the matrix. In the case of a filled system, it is easy to undersand, that if there are significant interactions between the resin and the surface then there is the possibility of an immobilized region of polymer in the interfacial region which will have a higher T_g than that of the bulk polymer. Similarly, if the filler reduces the overall density of the matrix material then there are likely to be regions which are more open than the average. It is however surprising that these regions are infact as large as they appear to be.

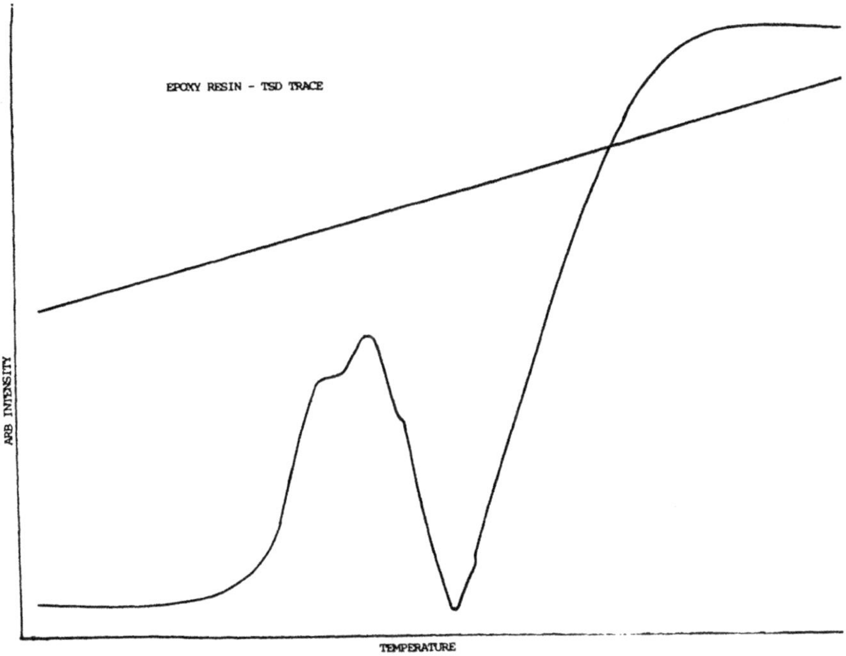

Figure (3) Multiple peaks observed in the TSD traces of an epoxy resin matrix containing a filler system

The technique allows one to observe the change in the position of the glass transition peak as a function of the extent of cure and hence to allow correlation with the degree of reaction, figure (5).

INTERFACIAL REGIONS IN COPOLYMER SYSTEMS

Polyurethanes are widely studied elastomers in which the crosslinks may either be generated by use of a tri functional reactive unit or as a result of the phase segregation of the copolymer and the subsequent hydrogen bonding. The dielectric studies of this system have revealed a number of interesting features.

To illustrate this type of system we shall consider the case of the MDI-polyethyleneadipate-MDI prepolymer chain extended with either 1,4 or 1,3 butanediol. In the case of the 1,4 butandiol the copolymer will tend to form well defined hard blocks which are fairly well phase separated. The addition of 1,3 butanediol causes a disruption of the hard block structure and as a consequence a greater intermixing of the soft and hard blocks with a consequent increase in the interactions and consequently the glass transition temperature. These trends are clearly illustrated in the case of mechanical studies of the copolymers figure (4). It is also possible to use the modulus to derive the effective crosslink density. The values correlate relatively well with those obtained from the swelling data and indicate that the crosslinking

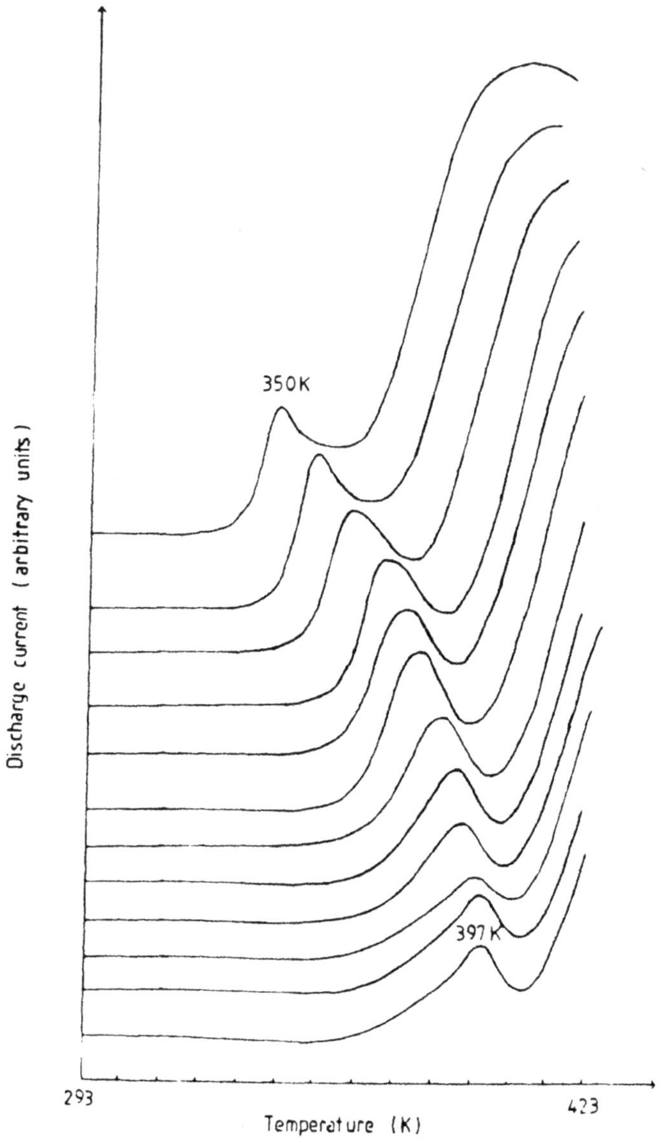

Figure (4) TSD traces illustrating the shift in the glass transition peak as a result of cure of the resin

which can be associated with the elastomeric properties is associated with a combination of urethane and chain entanglement interactions. A detailed study of the dielectric measurements on these systems also indicates that it is possible to observe two relaxation processes with slightly different activation energies. A detailed analysis of these

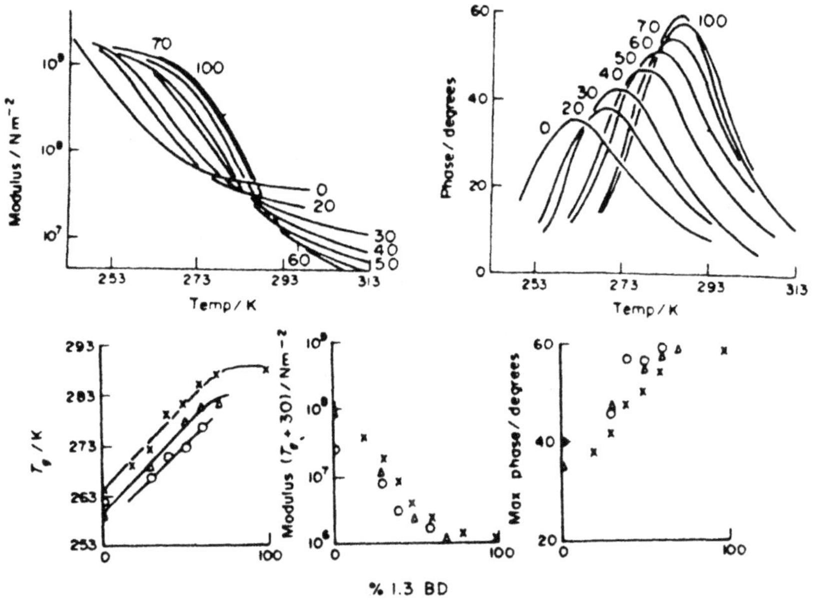

Figure (5) Mechanical relaxation spectra and analysis for the effects of the variation of 1,3 to 1,4 butanediol ratio on the properties of segmented polyurethane elastomers.

systems indicates that the two processes can be associated with relaxation of the dipoles in the completely phase separated region of the polymer and also in the interfacial region, figure (6).

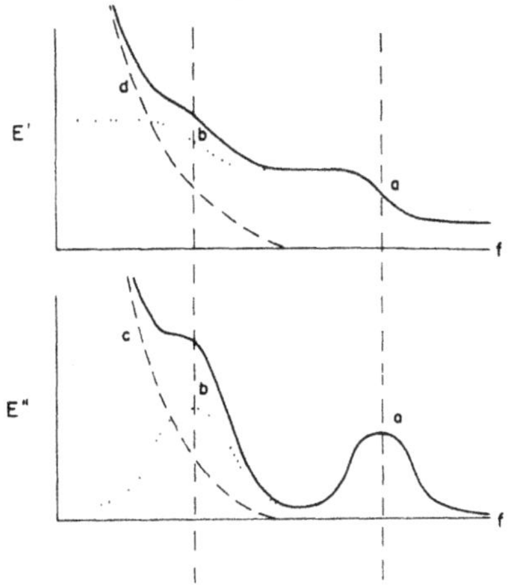

Figure (6) Dielectric relaxation curves for segmented polyurethanes.

The typical elastomer may therefore be modelled in the following way, figure (7).

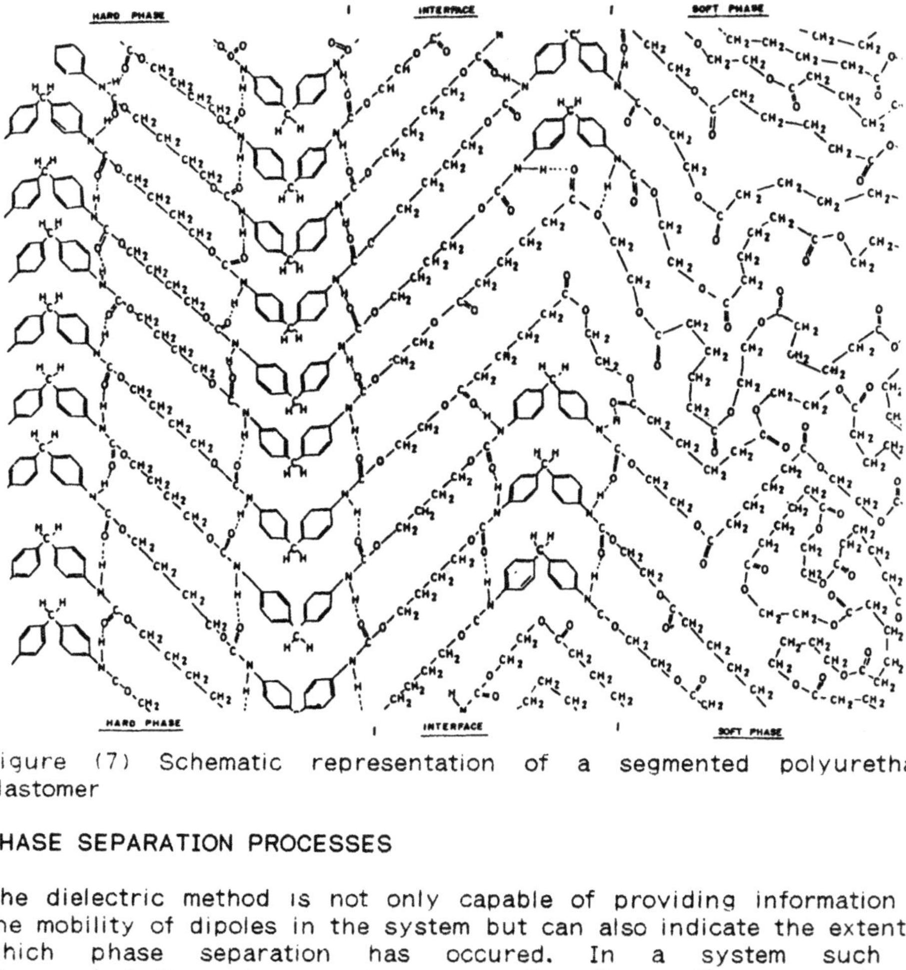

Figure (7) Schematic representation of a segmented polyurethane elastomer

PHASE SEPARATION PROCESSES

The dielectric method is not only capable of providing information on the mobility of dipoles in the system but can also indicate the extent to which phase separation has occured. In a system such as styrene-butadiene-styrene phase separation figure (8) can produce a well defined morphology which will conribute to the dielectric relaxation spectrum of this material. It is not possible to cover the complete theory of the so-called Maxwell Wagner Sillers process in the article, however it can be modelled using a knowledge of the length to breadth ratio of the microphase and the ratio of the conductivities of the two phases. In many copolymers, a low frequency dielectric process can be observed which is associated with the interfacial polarization process. The processes itself involves the transport and trapping of charge carriers to the interfacial region. Broadly speaking, the amplitude and

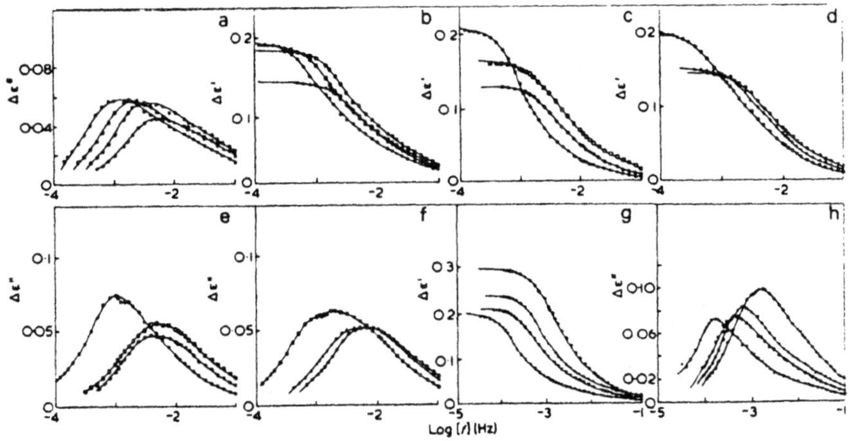

Figure(8) Low frequency dielectric relaxation in doped samples of styrene-butadiene-styrene triblock copolymers

time of the process are directly proportional to the distance vector in the direction of the field. In the context of curing systems examination of a novolac system during cure clearly indicates the point at which phase separation is detecte and this can be associated with a model of gelation in which the matrix is non-uniformly crosslinked and hence is an inhomogeneous structure.

CONCLUSION

Dielectric methods can provide useful information on both the local mobility of dipoles in the matrix and also on the extent to which phase separation has occured in the system. The method is however less useful for the estimation of the extent to which crosslinking has occured, however recently it has been shown that in certain systems it is possible to correlate changes in the dielectric spectrum with the reptation motion of polymer molecules and hence is in a way giving information on diffusive motions in the system.

Recommended reading

1)R.T.Bailey, A.M.North and R.A.Pethrick, Molecular Motion in High Polymers Oxford University Press (1983)
2)A.M.North R.A.Pethrick and A.D.Wilson Polymer **19**, 913 (1978)
3)L.Bunton,J.H.Daly and R.A.Pethrick J.Materials Science **18**, 2817 (1983)
4)R.A.Pethrick and A.R.Eastwood European Polymer J **19**, 43 (1983)
5)M.G.B.Mahboubian Jones D.Hayward and R.AS.Pethrick European Polymer J. 11 855 (1987)

ULTRASONIC AND VISCOELASTIC STUDIES OF CHAIN ENTANGLEMENT IN POLYMER SOLUTIONS

RICHARD A. PETHRICK
Department of Pure and Applied Chemistry
University of Strathclyde, Cathedral Street
Glasgow G1 1XL

INTRODUCTION

Whether a polymer chain forms a permanent or temporary entanglement or crosslink; one of the most drammatic effects which is observed is the modification of the dynamic spectrum exhibited by the polymer chain. The other manifestation of network formation is the observation of a modulus value which is directly related to the number of network points per unit volume. In general, we can divide networks into two types; firstly, those in which the chains between the network points are topographically defined in space - "rigid networks" and secondly, those that have flexible regions between junction points - "flexible networks". The most characteristic feature of the dynamics of a polymer chain is its glass transition temperature - T_g. Above T_g, elements of the polymer chain can undergo co-operative molecular motion, below Tg only the uncorrelated motion of short elements of the backbone and side chains are possible. The division of the networks into rigid and flexible can be made on the basis of the glass transition temperature of the backbone polymer structure. In order to understand how the network formation influences the dynamic properties of the matrix it is appropriate to review the measurements which have been made on isolated polymer chains and the way in which their properties are modified on the formation of three dimensional network structures.

DYNAMIC PROPERTIES OF ISOLATED POLYMER CHAINS IN SOLUTION

A polymer chain will adopt a statistical structure which depends on the nature of polymer solvent interactions. In a poor solvent system the chain will collapse and the nature of the interactions between neighbouring bonds will increase and this is reflected in the chain dynamics. For simplicity we will consider the dynamics of an ideal flexible chain in a good solvent. An example of such a system would be polydimethylsiloxane dissolved in toluene. A wide range of techniques exist for the study of the dynamic properties of polymer solutions and include; nuclear magnetic resource, fluorescence lifetime

O. Güven (ed.), Crosslinking and Scission in Polymers, 139–151.

measurements, dynamic light scattering, viscoelastic, ultrasonic and dielectric measurements. In this chapter we will consider the use of ultrasonic and viscoelastic measurements as a method of characterising chain dynamics.

ULTRASONIC RELAXATION MEASUREMENTS ON POLYMER SOLUTIONS

The frequency dependence of the alternation of high frequency sound waves has been used for the study of the dynamic properties of small and large molecules. Sound energy is transmitted through a liquid via the translational motion of molecules in the solution. A pressure wave can induce, by adiabatic compression of the media, an increase in the translational temperature of the solution. If the molecules undergo inelastic collisions it is possible to convert a proportion of the excess translational energy into rotational, vibrational or conformational energy. The dissipation of energy depends on the rate at which energy is returned from the excited state to the pool of translational energy. The rate at which this occurs is a characteristic of the natural relaxation rate of the system and is a function of the energy states of the molecular system. The ultrasonic attenuation (α/f^2) can be described as a function of frequency by the relation:

$$\frac{\alpha}{f^2} = \frac{A}{1+(f/f_c^2)} + B \qquad (1)$$

where A is the amplitude of the process, f_c is the characteristic frequency and B is, for a low molecular weight system the viscothermal loss, characterised by the Navier-Stokes equation which has the form:

$$B = \frac{4\pi^2}{\rho c^3}\left\{(\eta_s + {}^4/_3\eta_v) + \frac{K(\gamma-1)}{C_p}\right\} \qquad (2)$$

where η_v and η_s are respectively the shear and volume viscosities of the solution, ρ is the density, c is the velocity of sound, γ is the ratio of the specific heats; C_p at constant pressure to that at constant volume C_v and K is the thermal conductivity of the solutions. The sound attenuation coefficient (α/f^2) will depend on frequency for a system capable of undergoing conformational change, figure (1). As the temperature of the system is changed so the values of A and f will also vary. The temperature dependence of f_c is related to the relaxation rate (τ) by the relation $\tau=1/2\pi f_c$. The temperature dependence for a simple thermally activated process is described by an Arrehius relationship:-

$$\tau = A\exp(\Delta H^{\#}/kT)\exp(\Delta S^{\#}/k) \qquad (3)$$

where A is the pre-exponential factor and $\Delta H^{\#}$ and $\Delta S^{\#}$ are respectively the enthalpy and entropy of activation for the rotational isomeric process from the excited to the ground state. This approach has been used for the study of a wide range of substituted ethane molecules,

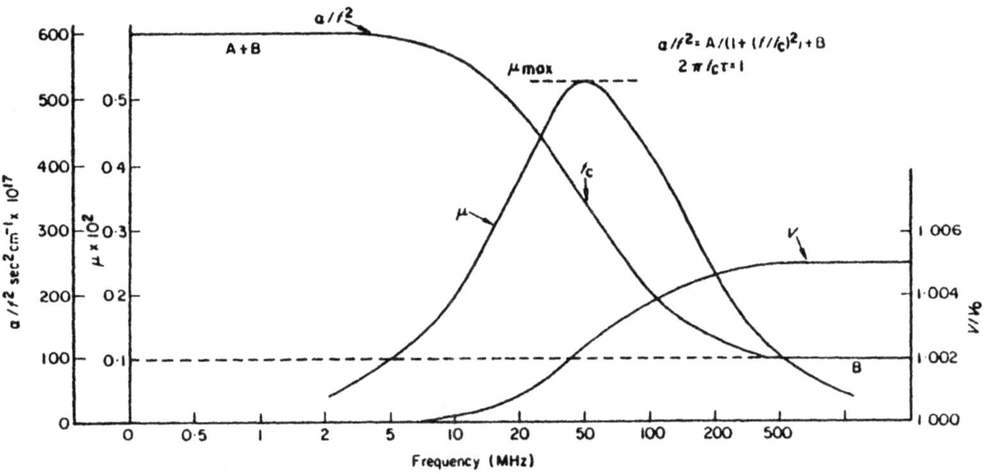

Figure (1) Frequency dependence of the sound propagation in the region of an ultrasonic relaxation

substituted propanes and butanes, cyclic substantiated structures, dioxanes, sulphites and sulphates. The study of polymer molecules in solution is however more complex and this is a direct consequence of the number of degrees of freedom for internal motion possible in a long chain structure. Studies of the activation energy for the processes vary with chain length, table (1). In the case of ethane, no energy

Table 1 Activation energies for internal rotation in hydrocarbons

Molecule	n	Activation energy, $H^{\#}$ (kJ.mole^{-1}) Internal rotation	Viscous flow
Ethane	2	13.4	-
Propane	3	14.8	-
Butane	4	14.6	-
Pentane	5	16.3	-
Hexane	6	14.2	8.0
Heptane	7	13.6	8.5
Octane	8	13.0	9.0
Nonane	10	13.8	10.9
Decane	11	11.2	11.9
Dodecane	12	15.5	14.0
Tetradecane	13	19.3	15.9

n.b. the first four molecules are gases at room temperature

difference exists between conformational states sound energy. Butane will have two iso energetic higher energy states when the molecule adopts a cis structure and a lower energy true state. Exchange between these structures will lead to the absorption or dissipation of energy and a consequent attenuation of the sound wave. The rate of exchange is a direct measure of the natural frequency of thermal deactivation of the excited state. Whilst the energy states of butane can be described in terms of three energy values, two of which are iso energetic; the addition of a further bond to create propane produces system which is complex. The conformational states can be described in terms of rotational angles associated with bonds two (Φ_2) and three (Φ_3) and lead to an energy surface. The minimum and maximum stable energy states will now be generated by the rotation of both (Φ_2) and (Φ_3), figure (2). The potential energy barrier is a function of the

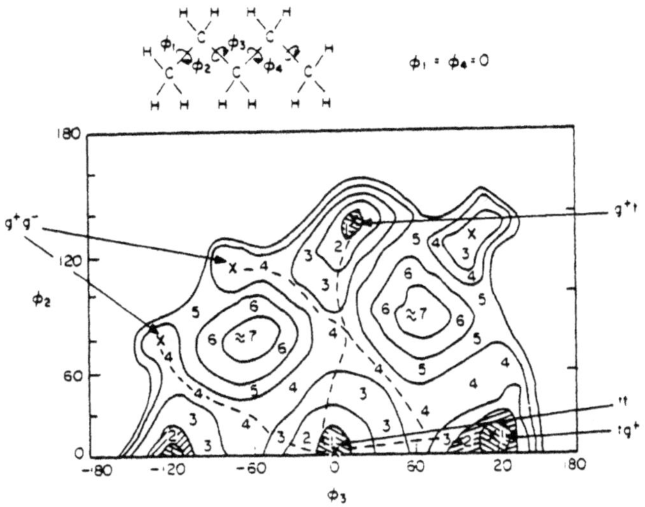

Figure (2) Potential energy contour diagram for pentane.
The terminal bonds are assumed to be fixed with Φ=0; the number indicate the most probable paths for the exchange between isomeric forms, trans = t, gauche = g, + and - refer to right and left handed rotations respectively.

environment created by surrounding molecules. A study of the effects of molecular weight on the amplitude and frequency of the relaxation of narrow molecular weight polystyrene in toluene exhibits a marked dependence on mass upto a value of 10^4 , figure (3). This observation can be explained on the basis of end group effects. The potential energy profile associated with the conformational changes of the bonds at the end of the chains are different and in this case lower than these associated with the centre of the polymer chain. At a molecular weight of about 10^4 the contribution from the ends of the polymer is small and it is only the relaxation of the centre chain which influences the ultrasonic relaxation. Careful analysis of the absorption curves indicate that in addition to the conformational relaxation, the attenuation through a frequency dependent component to B of equation (1).

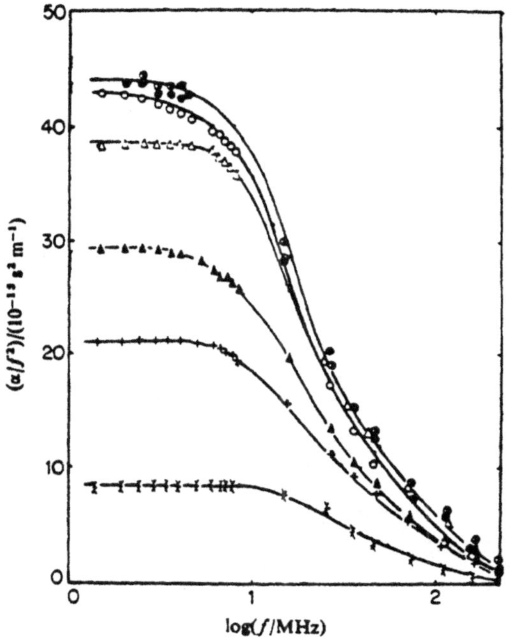

Figure (3a) Ultrasonic relaxation in narrow molecular weight samples of polystyrene in toluene

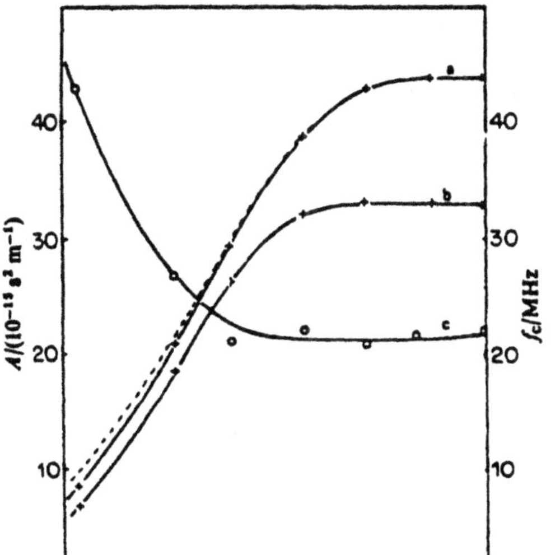

Figure(3b) Plot of the amplitude,characteristic frequency and reduced amplitude as a function of molecular weight for polystyrene

VISCOELASTIC PROCESSES

The frequeny dependence of the low viscosity of a polymer solution reflects the cooperative motions of the whole polymer chain. The first normal mode for the polymer chain is a cooperative distortion of the chain which generates a mode at the centre of the molecule.Higher normal modes have associated with them an increasing number of modes and there is a consequent reduction in the frequency dependent flow viscosity. The behaviour of an ideally flexible polymer chain is illustrated in figure (4). The tail of the normal mode processes

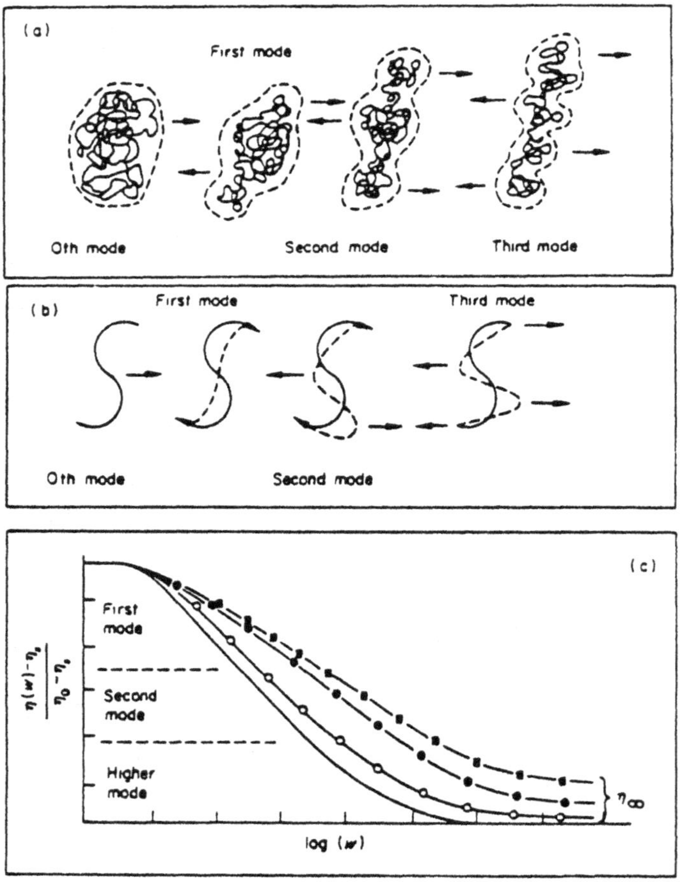

Figure (4) Schematic diagram of viscoelastic relaxation.

extends into the MHz frequency range and contribute a small, but significant, contribution to the ultrasonic attenuation. The mode distribution in an isolated polymer chain is purely a function of the normal mode spectrum and as such is only a property of the end-to-end distance of the random chain structure. In more concentrated solutions the effects of polymer-polymer chain contents have to be considered. Two factors should be considered; molecular weight of the polymer and the intrinsic backbone flexibility. In the ideal case, the ultrasonic attenuation, corrected for the effects of normal mode relaxation should be independent of concentration, i.e., the attenuation directed by the segment density should be a constant. Studies of polystyrene solutions have shown that this is not the case and have lead to speculations about the origins of these derivations. In order to understand these effects it is useful to examine the viscoelastic properties of polydimethyl siloxane -PDMS. A study of the viscosity of PDMS as a formation of molecular mass indicates that the polymer behaviour can be divided into two regions. Low molecular weights, values below about 14,000, exhibit a viscosity - molar mass dependence of the form: η=KM, figure 5. Above 14,000 a higher dependence of the molecular weight is observed;η=KM$^{3.5}$.

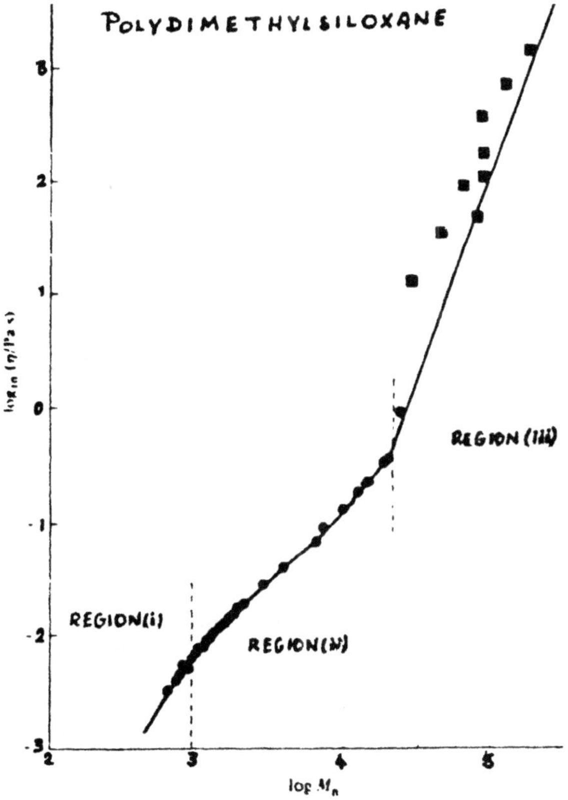

Figure(5) Variation of viscosity with molecular weight for polydimethlsiloxane

This higher order coefficient is believed to be a consequence of polymer-chain entanglements. In the model of Doi and Edwards it is proposed that the long time-low frequency viscoelastic relaxation of the melt is a consequence of the reptational motion of a polymer chain through a tube formed from the entanglements created by the surrounding polymer chains. The rate of the relaxation will depend on the number of entanglements and these will in turn also dictate the magnitude of the observed viscosity component. The classical Doi Edwards theory only considers the contributions to the viscosity arising from the reptation motion and a full explanation of the observed behaviour requires the addition of two further contributions. The first of these is the contribution due to the cooperative normal mode motions of the lengths of chain between entanglements and the final contribution is due to the resistance generated by the freezing out of the local motions of short elements of the backbone.

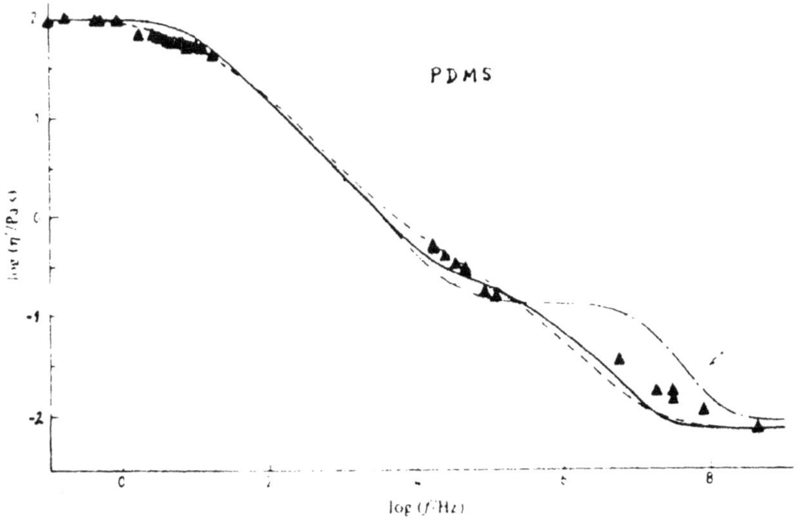

Figure (6) Viscoelastic relaxation in linear polydimethysiloxane.

These three regions of the viscoelastic spectrum can be observed experimentally, figure (6) and will introduce a source of frequency dependence to the ultrasonic relaxation spectrum. Studies of the viscoelastic properties of PDMS have shown the essential correctness of the "reptation" model but have also highlighted the fact that the basic Doi Edwards theory requires modifications to allow for the molecular mass distribution before it is possible to attain an acceptable fit of the experimental data.

ULTRASONIC STUDIES OF ENTANGLED POLYMER SYSTEMS

An increase in the molar mass of a polymer will be expected to increase the normal mode relaxation time and hence shift this relaxation contribution to lower frequencies. Studies of polystyrene have supported this analysis. If however concentrated solutions or melts are used then other effects can be identified. In the case of PDMS, the local sequential motions of the backbone will occur at frequencies above 1GHz and hence the ultrasonic relaxation is purely a result of the normal mode processes. Initially an increase in molar mass leads to shift in the spectrum to lower frequency, however, around the values for critical entanglement an additional contribution to the relaxation spectrum is observed, figure (7).

A theoretical analysis of the dynamic spectrum indicates that the observed spectrum can be described in terms of a super position of two features. The two entangled chains will have an effective molecular weight corresponding to the combined value of the chains minus the free ends.

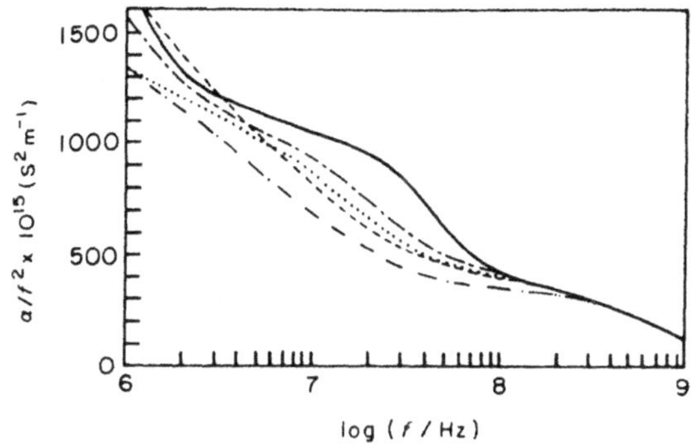

Figure (7a) Ultrasonic relaxation curves for linear polydimethylsiloxane

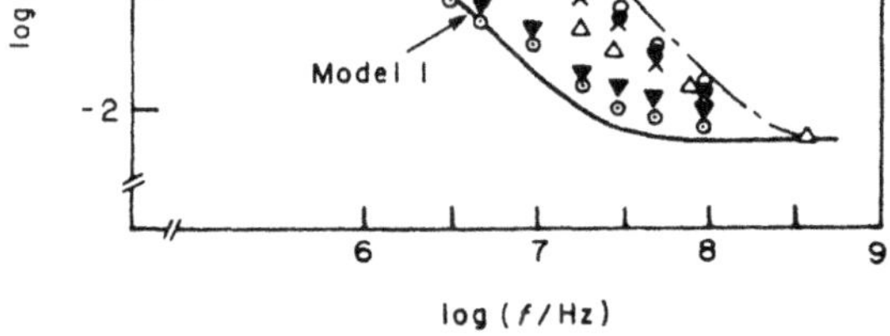

Figure (7b) Dynamic shear viscosity for linear polydimethylsiloxanes

The entanglement will also dictate the mode spectrum and the first normal mode and subsequent mode spectra will have to exhibit a node at the entanglement point. The "ends" of the chains will however be able to exhibit normal mode spectra characteristic of these low molecular weight free ends, figure (8).

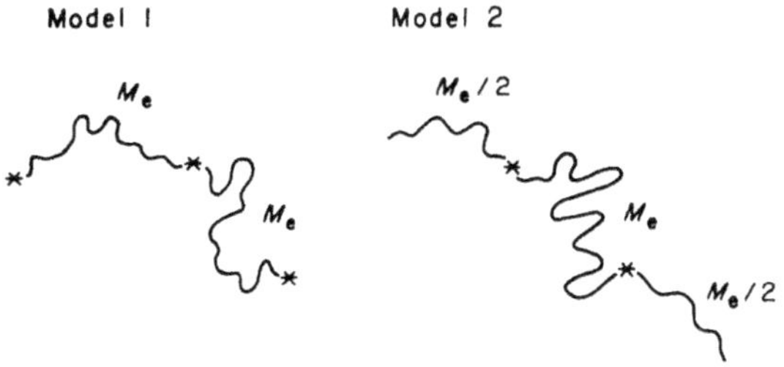

Figure(8) Representation of the mode structures for entangled polymers

This latter motion explains the observation of the additional contribution figure (7) and accounts for the molar mass sensitvity. Once the critical entanglement value has been achieved the main part of the mode spectrum is shifted to lower frequencies and the average value between entanglements becomes constant. This latter process gives an apparently molecular weight independent contribution to the attenuation. The effects of end groups will only be observed for chains with molar mass close to the critical value and as the "effective" molecular mass of the entangled matrix increases, the contribution from the end groups will rapidly decrease. Comparison of viscoelastic and ultrasonic measurements have unambiguously demonstrated the validity of this description of polymer chain dynamics. These experiments illustrate the way in which the creation of a physical entanglement or network point leads to a modification of the relaxation spectrum exhibited by an ideal polymer chain. The same processes will be expected to occur when a polymer chain is crosslinked and are similar to those observed in the case of the ideal isolated flexible chain i.e., the generation of a high value low frequency component characteristic of the overall matrix and increase of the high frequency spectrum corresponding to cooperative motions of the chain elements between entanglement points. There is an additional effect of the formation of interaction points in the matrix which is often not recognised and this is illustrated by a recent study of the process of phase separation in styrene-butadiene-styrene triblock copolymer. -(SBS).

PHASE SEPARATION IN SBS SOLUTIONS

The relaxation spectra of SBS triblock copolymer in solution may be described in terms of the superposition of contributions to the relaxation spectrum from processes associated with motion of the styrene and butadiene blocks and normal mode relaxation of the whole polymer chain. At the point at which the triblock copolymer phase separates to form a three dimensional network is identified by the observation of a marked increase in the high frequency attentuation coefficient as the temperature decreases, figure (9). This contribution is not a result of scattering of the sound wave by the islands of styrene but is a direct result of the generation of regions of dense polymer structure which is impermeable to the flow of solvent molecules. The styrene domains become like billiard balls joined by permeable elastic structures and the effect is to create an additional component ot the viscosity term of the Navier-Stokes equation which will become frequency dependent at a value close to that for the solvent. In a network we may anticipate that there should exist a high frequency component associated with the rigid network structure of the matrix.

Ultrasonic studies have been reported recently on the changes which occur in the relaxation spectrum as an aliphatic isocyanate is cured with a reactive diol-triol mixture. Initially the ultrasonic attenuation is relatively low, however as the molecular weight of the structure increases so a significant contribution to the attenuation due to normal mode relaxation can be identified.

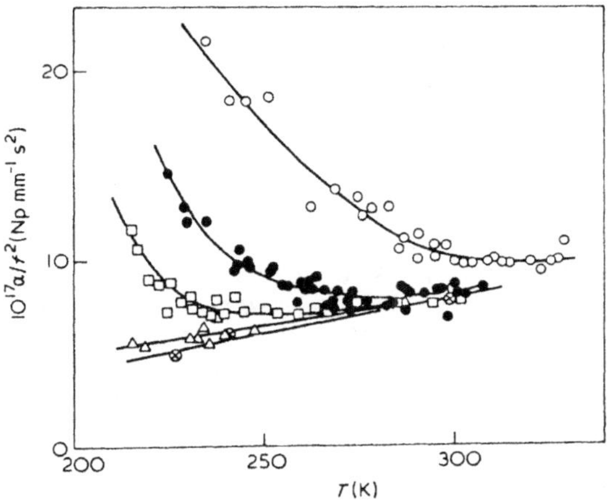

Figure (9) Temperature dependence of attenuation at 740 MHz for SBS solutions in toluene

Since the relaxing unit corresponding to the chain extended isocyanate forms into a gel by the incorporation of the triol, the size of the normal mode is relatively large in these systems and hence the contribution to the ultrasonic attenuation in the MHz range. It is therefore possible, in principle, to correlate the ultrasonic attenuation with the extent of network formation. These studies on flexible gel structures are still in their infancy.

ULTRASONIC STUDIES OF NETWORK FORMATION IN RIGID NETWORKS

The study of rigid networks has been very limited and this is probably a consequence of the difficulty of these measururements. Two systems; novolacs and epoxy resins have however been studied to a limited extent.

i) NOVOLAC RESINS

Novalacs resins have typically a molecular mass of the order of a couple of thousand and are cured by the addition of a methylating-crosslinking agent such as tetra methylene tetra amine. If the measurements are carried out at the temperature of the core characteristic changes are observed in both the attenuation and velocity of sound. The initial increase in the attenuation can be associated with a coincidence of the frequency of local motion of the chain and also normal mode motion with the observation frequency. As the cure process proceeds so the length of the elements COVALENTLY linked increases, but this is modulated by the formation of the three dimensional network. As cure proceeds the rigidity of the matrix increases and this decreases the amplitude of the segmental rotation with a consequent desreux in the ultrasonic attenuation. As the matrix approaches the gel point then the overall rigidity increases markedly with a corresponding increase in the sound velocity. The compressibility of the matrix can be simply related to the sound velocity by the relationship:-

$$K = 1/\rho c^3 \qquad -(4)$$

where K is the adiabatic compressibility, ρ is the density and c is the sound velocity. Initially the matrix is fluid, has a high compressibility and a velocity in the region of 1400 m/sec. Cure decreases the compressibility and raises the sound velocity to a value of approximately 3000 m/sec. The point at which the attenuation changes will not coincide with that for the velocity as they sense different aspects of the system. The attenuation contains information on both the mode structure of the matrix and also the extent to which the backbone can undergo internal rotation. The velocity changes associated with the variation of the molecular motion is small compared with the much larger changes which occur between the fluid and rigid states.

ii) EPOXY RESINS

A study of solid epoxy resins has been performed in which cure is produced by reaction of the diglycydl ether of bisphenol A with a solid diamine. The matrix is solid at the start and cure does not change the state of matter and hence the changes in the velocity are not very large. Similarly, since the attenuation is dominated by the local motion of the backbone, the changes which occur are also not very marked. The initial decrease in the velocity can be associated with changes which occur in the matrix during cure. The density decreases as a result of the generation of voids in the matrix and this also leads to a reduction in the sound velocity. As the cure proceeds so these voids are eventually removed and the velocity is once more observed to increase as a homogeneous solid matrix is generated.

In summary, ultrasonics is useful for the study of cure and network formation in flexible systems but is of limited application in the characterisation of solid matrices.

RECOMMENDED READING

1 RA Pethrick, Ultrasonic Studies of Macromolecules, Prog Polym Sci **9** 197-295 (1983)

2 JD Ferry Viscoelastic Properties of Macromolecules Interscience New York London 1978

3 RT Bailey AM North and RA Pethrick Molecular Motion in High Polymers Pergamon Oxford 1981

SPECIFIC PAPERS

1 RA Pethrick and AM Abubaker Polymer Communications **27** 194-196 1986

2 RA Pethrick RR Rahalker J Lamb G Harrison AJ Barlow W Hawthorne JA Semlyen and AM North Proc Roy Soc London **A394** 207-222 1984

DETERMINATION OF CROSS-LINK DENSITY IN AMORPHOUS NETWORKS BY STRESS-STRAIN-SWELLING EXPERIMENTS

Burak Erman
Polymer Research Center
Boğaziçi University
Bebek, 80815
İstanbul, Turkey

ABSTRACT. The degree of cross-linking in an amorphous elastomeric network is related to the shear modulus which may be determined either by stress-strain experiments or by equilibrium swelling measurements in a solvent. The value of the cross-link density determined in this manner depends on the molecular model used. In the present paper, the molecular constitution of affine and phantom models is reviewed in detail, followed by a discussion of the constrained junction model of real networks. Results of experiments are compared with predictions of the theory and an improved method of determining cross-link density is discussed.

1. INTRODUCTION

The determination of cross-link density by stress-strain experiments or by equilibrium swelling measurements is well-known and has been discussed in various books (1-3). The value of the cross-link density calculated in this manner depends on the molecular model chosen. The simplest and most commonly used molecular models are the affine network model of Wall and Flory (4), and the phantom network model of James and Guth (5,6). Experimental work on various aspects of rubbers since the dawn of these two theories has shown, however, that a network tested in the dry state conforms neither to the phantom nor to the affine model. Experiments in simple tension, in particular, show that the modulus of a network is closer to that of the affine model for small deformations, and approach to that of the phantom model with increasing strain. Similarly, the modulus of a network tested in the swollen state is not equal to that of the dry network, the former being significantly less than that of the latter.

Various improvements in the theory for understanding the strain dependence of the modulus of a network have been proposed after 1975. The original idea describing the decrease of the modulus upon stretching is given by Ronca and Allegra (7). According to the theory proposed by these authors, the constraints affecting the fluctuations of junctions are more effective at smaller strains and become progressively less important as extension increases. In a different approach, Deam and

O. Güven (ed.), Crosslinking and Scission in Polymers, 153–169.

Edwards (8) have introduced the effect of topological entanglements into network behavior. The theory based on this paper was later presented in terms of the slip-link model (9,10) according to which the strength of entanglements surrounding a given network chain depends on the imposed macroscopic strain. According to this treatment, the small strain modulus shows contributions from entanglements trapped permanently into the network structure by cross-linking. Such effects are modelled by links joining different chains. At higher deformations, contributions from such entanglements vanish and the modulus becomes proportional to the number of chemical cross-links. The ideas proposed by Ronca and Allegra (7) have been put into a mathematically and physically more rigorous form by Flory (11). According to this model, which has subsequently been termed as the 'constrained junction model', the intermolecular constraints or entanglements acting at all points along the chains are assumed to be concentrated at the junctions. These constraints depend on strain and swelling. Their effects diminish with increasing strain and swelling, leading to a substantial decrease in the modulus at high extensions and/or swelling. Unlike the model presented by Edwards, trapped entanglements do not contribute to the small strain modulus in the constrained junction model (see sequel). The constrained junction model has later been improved (12,13) and quantitative agreement has been obtained between the predictions of the theory and results of stress-strain experiments on various elastomeric networks.

The molecular models cited in the preceding paragraph, as well as various extensions of these by various authors have been reviewed by Eichinger (14) without going into mathematical details. Discussion of various elasticity theories may also be found in the monograph by Mark and Erman (3).

All of the modern theories cited above indicate that a state of phantom-like network behavior is obtained at high degrees of extension and swelling. The phantom network model thus constitutes a convenient reference structure on which the behavior of real networks may be superposed.

In the present paper, the molecular constitution of a network is defined, followed by a discussion of the affine, phantom and constrained junction models. Predictions of the moduli of networks based on these models are compared with results of experiments and a method of estimating the correct cross-link density is suggested.

2. MOLECULAR CONSTITUTION OF A NETWORK

An amorphous elastomeric network consists of long flexible chains joined together to form a three dimensional solid structure. The chains are joined together by permanent chemical bonds either randomly located along their lengths or at specific locations. This method of joining the chains together is called cross-linking. Alternatively, the chains may be joined together only at their ends, called end-linking. The numbers of chains and junctions in the network are denoted by ν and μ, respectively. The functionality ϕ of a junction is defined as the number of chains meeting at that junction. Some of the chains may be attached to only one junction. Such chains are called dangling chains and do not contribute to the elastic activity of the network. Similarly, a loop which is a chain starting and terminating at the same junction does not contribute to elastic activity of a network. A junction of functionality 2 does not contribute to the elastic activity of the network but causes only chain extension. A network in which there are no loops, dangling chains and $\phi \geq 3$ for all junctions is called a perfect network. The cycle rank ξ of a network is defined as the number of independent circuits, or the number of cuts required to reduce the connected structure to a tree. The cycle rank is a measure of the connectivity of a perfect as well as an imperfect network. The structure of networks is discussed in detail by Flory (15). For a perfect network, the parameters μ, ν, ϕ and ξ are related by two equations,

$$\mu = 2\nu/\phi \qquad\qquad (1)$$
$$\xi = (1 - 2/\phi)\nu$$

Denoting the average molecular weight of chains between junctions in a network by M_c, another expression may be written as

$$\xi/V_o = (1 - 2/\phi)\rho N_A/M_c \qquad\qquad (2)$$

where V_o is the volume of network during formation, ρ is the corresponding density and N_A is the Avogadro number. Modification of eqs 1 and 2 for imperfect networks requires further discussion which will not be undertaken in the present review. The reader may refer to references 1 and 3 and papers by Scanlan (16), Case (17) and Flory (18) for imperfect network.

As will be shown below, contributions to the modulus of a real network do not contain effects from dangling chains and loops. The experimentally obtained modulus is a direct measure of the cycle rank or of connectivity. Stress-strain-swelling experiments thus lead to information on the values of μ and ν of the perfect network only. Without further information from an independent source, stoichiometry for example, the numbers of inactive chains that do not participate in the elastic activity of the network cannot be estimated.

3. MACROSCOPIC DEFORMATION

The molecular theory of rubber elasticity requires proper use of the measure of macroscopic deformation. In this section the state of macroscopic deformation is outlined for the two widely employed experiments, i. the uniaxial (or simple) tension and ii. homogeneous swelling.

i. Simple Tension. Dimensions of a test specimen in the reference and deformed states are shown in Figure 1. L_o in Figure 1a indicates the length of the specimen in the reference state which may conveniently be taken as the state of formation of the network. In the case of network

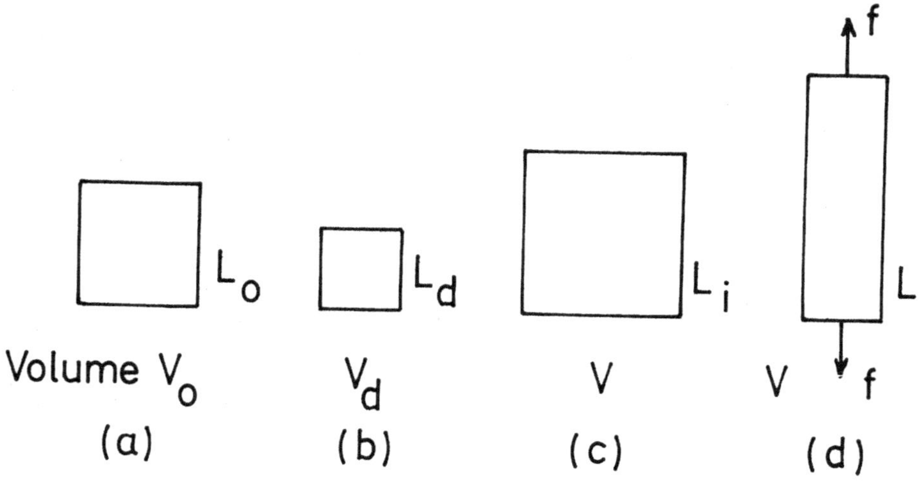

Figure 1. Dimensions of a test specimen in simple (uniaxial) tension. (a) The specimen in the state of formation, which may be taken as reference stress-free state. L_o and V_o are the reference length and volume, respectively. (b) The dry state of the network, L_d and V_d denoting the dry length and volume, respectively. (c) The initial swollen state of the network. For experiments in the swollen state, the network is swollen with a sufficiently nonvolatile solvent and stretched in air. V is the volume of network at the start of the experiment, which is assumed to be constant during stretching. L_i is the initial swollen but undistorted length. (d) The length L obtained under the uniaxial force f

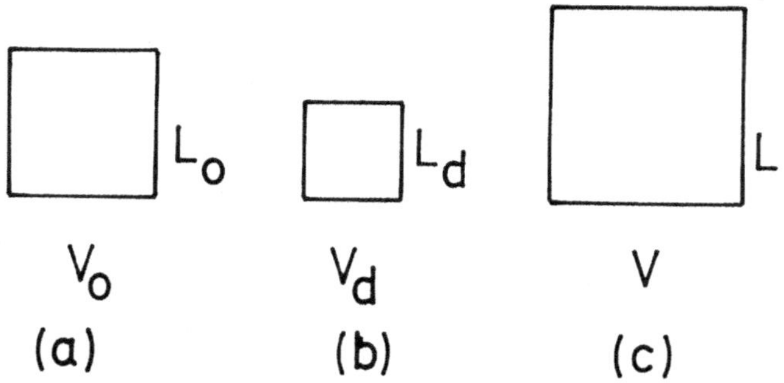

Figure 2. Dimension of the network in swelling experiments. (a) The state of the network during formation. V_o is the stress-free reference volume. (b) The dry state, represented by the volume V_d of the dry bulk network. (c) The final swollen volume V.

formation in the presence of a solvent, the total volume V_o of the network is comprised of the volumes V_s of solvent and V_d of the dry polymer. Figure 1b shows the length L_d of the dry network of volume V_d. Figure 1c depicts the dimensions of the network which has subsequently been swollen with a suitable (sufficiently nonvolatile) solvent. In this case the length has changed to L_i. This is the state of the network at the start of the experiment. The subscript i refers to the initial state where the volume is V. The change of dimensions from Figure 1a to 1b to 1c are isotropic. If the simple tension experiment is to be performed in the unswollen state, then $L_i = L_d$. Figure 1d depicts the final shape with length L obtained under a force f along the x-direction. The experiment is performed in air or preferably in an inert atmosphere. The volume change is negligibly small during stretching. The final volume in the stretched state may therefore be identified with the volume V of the swollen undistorted state. The volume fraction v_{20} of polymer during the formation of network is defined as

$$v_{20} = V_d/V_o = (L_d/L_o)^3 \qquad (3)$$

Once the network is formed, v_{20} is fixed. The volume fraction of polymer v_2 during the experiment is defined as

$$v_2 = V_d/V = (L_d/L_i)^3 \qquad (4)$$

v_2 may be changed by removing or adding solvent into the network prior to extension. Constant v_2 data for the stress-strain relations requires the use of a sufficiently nonvolatile solvent. Homogeneous diffusion of the solvent into the network before the stress-strain experiment is required.

The deformation gradient tensor $\underset{\sim}{\lambda}$ for the simple tension shown in Figure 1d is given by

$$\underset{\sim}{\lambda} = \begin{vmatrix} \lambda_x & & \\ & \lambda_y & \\ & & \lambda_z \end{vmatrix} = \begin{vmatrix} L_x/L_{ox} & & \\ & L_y/L_{oy} & \\ & & L_z/L_{oz} \end{vmatrix} \qquad (5)$$

where L_x, L_y and L_z are the dimensions of the sample in the deformed state along the x, y and z directions, respectively. L_{ox}, L_{oy} and L_{oz} are the corresponding components in the reference state. Components of the deformation gradient tensor defined in this manner thus relate the final dimensions of the sample to its dimensions in the state of formation. Another measure of deformation, the extension $\underset{\sim}{\alpha}$, is defined as

$$\underset{\sim}{\alpha} = \begin{vmatrix} \alpha_x & & \\ & \alpha_y & \\ & & \alpha_z \end{vmatrix} = \begin{vmatrix} L_x/L_{ix} & & \\ & L_y/L_{iy} & \\ & & L_z/L_{iz} \end{vmatrix} \qquad (6)$$

where L_{ix}, L_{iy} and L_{iz} are the dimensions in the swollen but undistorted state of the network. Experimental data are reported, in general, in terms of the components of $\underset{\sim}{\alpha}$. Using eqs 3 and 4, the i th component λ_i may be related to α_i as

$$\lambda_i = (v_{20}/v_2)^{1/3}\alpha_i \tag{7}$$

Using the definitons for $\underset{\sim}{\lambda}$ and $\underset{\sim}{\alpha}$, the following identities are obtained

$$\begin{aligned} \lambda_x\lambda_y\lambda_z &= V/V_o = (v_{20}/v_2) \\ \alpha_x\alpha_y\alpha_z &= 1 \end{aligned} \tag{8}$$

For uniaxial tension along the x direction, letting $\lambda_x = \lambda$ and $\alpha_x = \alpha$, one obtains

$$\begin{aligned} \underset{\sim}{\lambda} &= \begin{vmatrix} \lambda & & \\ & (v_{20}/v_2)^{1/2}\lambda^{-1/2} & \\ & & (v_{20}/v_2)^{1/2}\lambda^{-1/2} \end{vmatrix} \\ &= (v_{20}/v_2)^{1/3} \begin{vmatrix} \alpha & & \\ & \alpha^{-1/2} & \\ & & \alpha^{-1/2} \end{vmatrix} \end{aligned} \tag{9}$$

ii. *Swelling*. The dimensions of a sample during formation and in the swollen state are shown in Figure 2. Figure 2a depicts the volume V_o during formation. Figure 2b is the dry state and Figure 2c corresponds to the final swollen state of volume V. Inasmuch as isotropic swelling does not produce distortion of the sample, components of $\underset{\sim}{\alpha}$ equate to unity by definition and one obtains from eq 9,

$$\underset{\sim}{\lambda} = \begin{vmatrix} (v_{20}/v_2)^{1/3} & & \\ & (v_{20}/v_2)^{1/3} & \\ & & (v_{20}/v_2)^{1/3} \end{vmatrix} \tag{10}$$

4. MICROSCOPIC DEFORMATION

The state of macroscopic deformation in a homogeneously deformed network may be determined unambiguously by measuring the macroscopic dimensions before and after the application of the strain. The microscopic state of deformation, on the other hand, requires averaging over all the chains of the network before and after deformation. The mean-squared end-to-end chain length $\langle r^2\rangle_o$ between junctions averaged over all chains of the network in the reference state denoted by the subscript zero is assumed to be equal to the mean squared end-to-end distance of a single chain in the unperturbed state. Inasmuch as network formation is carried out in the isotropic state, the x, y and z components of $\langle r^2\rangle_o$ are equal and

$$\langle x^2\rangle_o = \langle y^2\rangle_o = \langle z^2\rangle_o = \langle r^2\rangle_o/3 \tag{11}$$

In the deformed state, the components $\langle x^2\rangle$, $\langle y^2\rangle$ and $\langle z^2\rangle$ change anisotropically, and the microscopic state of deformation may be defined in the mean-squared sense by the diagonal tensor

$$\underset{\sim}{\Lambda}^2 = \begin{vmatrix} \Lambda_x^2 & & \\ & \Lambda_y^2 & \\ & & \Lambda_z^2 \end{vmatrix} = \begin{vmatrix} \langle x^2\rangle/\langle x^2\rangle_o & & \\ & \langle y^2\rangle/\langle y^2\rangle_o & \\ & & \langle z^2\rangle/\langle z^2\rangle_o \end{vmatrix} \tag{12}$$

The relationship of the components of microscopic deformation tensor to the components of the macroscopic deformation requires a molecular model as will be described below.

5. THE ELASTIC FREE ENERGY OF THE SINGLE CHAIN AND THE NETWORK

The length of chains between cross-links in a typical network are sufficiently long to be considered Gaussian. Calculations by the rotational isomeric state sheme (19) indicate that the end-to-end chain vector distribution becomes approximately Gaussian for sufficiently flexible chains with more than 40-60 repeat units. Inasmuch as chains in a typical network are longer, they may satisfactorily be represented by the Gaussian distribution

$$W(r) = (3/2\pi\langle r^2\rangle_o)^{3/2}\exp(-3r^2/2\langle r^2\rangle_o) \tag{13}$$

where W(r) is the distribution of the end-to-end vector, $\underset{\sim}{r}$, and $\langle r^2\rangle_o$ is the mean-squared unperturbed value for the same chain. The elastic free energy A_{el} of the single chain whose ends are kept at a fixed distance r is given as (19)

$$\begin{aligned} A_{el} &= -kT\ \ln W(r) \\ &= C(T) + (3kT/2\langle r^2\rangle_o)r^2 \end{aligned} \tag{14}$$

where k is the Boltzmann constant, T is the temperature and the second line is obtained by substitution from eq 13. C(T) is a function of temperature. The average force f required to keep the chain ends at the fixed separation r is obtained from thermodynamics as

$$f = \partial A_{el}/\partial r$$
$$= (3kT/\langle r^2\rangle_o)r \qquad (15)$$

where the second line is obtained by substituting from eq 14. The term in parenthesis in eq 15 may be interpreted as a spring constant. Thus eq 15 shows that a Gaussian chain behaves as a linear spring.

The fundamental assumption of both the affine and phantom network models is that the total elastic free energy of a network is equal to the simple sum of the free energies of the constituent chains. Thus the elastic free energy ΔA_{el} of the network relative to the reference state is obtained by summing the free energies of the single chains as

$$\Delta A_{el} = \sum_{\nu}(3kT/2\langle r^2\rangle_o)r^2 - (A_{el})_o$$
$$= \frac{3\nu kT}{2}\left(\frac{\langle r^2\rangle}{\langle r^2\rangle_o} - 1\right) \qquad (16)$$
$$= \frac{\nu kT}{2}\left(\frac{\langle x^2\rangle}{\langle x^2\rangle_o} + \frac{\langle y^2\rangle}{\langle y^2\rangle_o} + \frac{\langle z^2\rangle}{\langle z^2\rangle_o} - 3\right)$$

Equation 16 is the starting point of the phantom and affine models. The third of eq 16 represents the elastic free energy of the network in terms of the components of the molecular deformation tensor $\underset{\sim}{\Lambda}^2$ defined by eq 12. For the complete formulation of the problem, the relationship of the components of $\underset{\sim}{\Lambda}^2$ to the macroscopic state of deformation has to be known. The form of this relationship depends on the molecular model chosen. The affine and the phantom models to be treated below are the two models that lead to simple expressions for ΔA_{el}.

6. THE AFFINE NETWORK MODEL

The affine network model assumes that the junction points are rigidly embedded in the medium, thus performing affine displacements with macroscopic strain. The components $\langle x^2\rangle$, $\langle y^2\rangle$ and $\langle z^2\rangle$ in the deformed state are related to those in the undeformed state as

$$\langle x^2\rangle = \lambda_x^2\langle x^2\rangle_o \ , \ \langle y^2\rangle = \lambda_y^2\langle y^2\rangle_o \ , \ \langle z^2\rangle = \lambda_z^2\langle z^2\rangle_o \qquad (17)$$

Substituting these expressions into eq 16 then leads to the elastic free energy of the network as

$$\Delta A_{el} = \frac{\nu kT}{2}(\lambda_x^2 + \lambda_y^2 + \lambda_z^2 - 3) \qquad (18)$$

A more general derivation of the affine network elastic energy by Flory (1) contains an additional logarithmic term that results from the distribution of cross-links over the sample volume. Thus the more complete elastic energy reads as

$$\Delta A_{el} = \frac{\nu kT}{2}(\lambda_x^2 + \lambda_y^2 + \lambda_z^2 - 3) - \mu kT \ln(V/V_o) \qquad (19)$$

The last term vanishes for a network whose final volume equals to that of the reference state.

7. THE PHANTOM NETWORK MODEL

In the phantom network model picture of James and Guth (5,6), the instantaneous values of end-to-end vectors for each chain may be written as the sum of a mean chain vector and fluctuation from it as

$$\begin{aligned} x_i &= \bar{x}_i + \Delta x_i \\ y_i &= \bar{y}_i + \Delta y_i \\ z_i &= \bar{z}_i + \Delta z_i \end{aligned} \qquad (20)$$

Here, the subscript i denotes the i th chain and x_i, y_i, z_i are the three components for the i th chain $\bar{x}_i$, $\bar{y}_i$, $\bar{z}_i$ and Δx_i, Δy_i, Δz_i denote the mean values and fluctuations from these means, respectively.

According to the phantom network model, mean chain vectors transform affinely with macroscopic deformation. The fluctuations, however, are not affected by macroscopic deformation. Thus eq 20 may be written in terms of undeformed values as

$$\begin{aligned} x_i &= \lambda_x \bar{x}_{i,o} + \Delta x_{i,o} \\ y_i &= \lambda_y \bar{y}_{i,o} + \Delta y_{i,o} \\ z_i &= \lambda_z \bar{z}_{i,o} + \Delta z_{i,o} \end{aligned} \qquad (21)$$

The subscript zero in eq 21 denotes the undeformed state. Squaring both sides of eqs 21 and taking the ensemble average, one obtains

$$\begin{aligned} \langle x^2 \rangle &= \lambda_x^2 \langle \bar{x}^2 \rangle + \langle (\Delta x)^2 \rangle \\ \langle y^2 \rangle &= \lambda_y^2 \langle \bar{y}^2 \rangle + \langle (\Delta y)^2 \rangle \\ \langle z^2 \rangle &= \lambda_z^2 \langle \bar{z}^2 \rangle + \langle (\Delta z)^2 \rangle \end{aligned} \qquad (22)$$

where the angular brackets denote an average over all the chains. The mean-squared values $\langle \bar{x}^2 \rangle$, etc. and fluctuations $\langle (\Delta x)^2 \rangle$, etc. are related to the chain vector components, $\langle x^2 \rangle$, etc. by the following relations (see, for example, reference 3 and the further reference cited therein)

$$\begin{aligned} \langle \bar{x}^2 \rangle_o &= (1 - 2/\phi)\langle x^2 \rangle_o \\ \langle \bar{y}^2 \rangle_o &= (1 - 2/\phi)\langle y^2 \rangle_o \\ \langle \bar{z}^2 \rangle_o &= (1 - 2/\phi)\langle z^2 \rangle_o \end{aligned} \qquad (23)$$

$$\langle(\Delta x)^2\rangle_o = (2/\phi)\langle x^2\rangle_o$$
$$\langle(\Delta y)^2\rangle_o = (2/\phi)\langle y^2\rangle_o \qquad (24)$$
$$\langle(\Delta z)^2\rangle_o = (2/\phi)\langle z^2\rangle_o$$

Substituting eqs 23 and 24 into eq 22 leads to the components of molecular deformation tensor as

$$\Lambda_x^2 = \langle x^2\rangle/\langle x^2\rangle_o = (1 - 2/\phi)\lambda_x^2 + 2/\phi$$
$$\Lambda_y^2 = \langle y^2\rangle/\langle y^2\rangle_o = (1 - 2/\phi)\lambda_y^2 + 2/\phi \qquad (25)$$
$$\Lambda_z^2 = \langle z^2\rangle/\langle z^2\rangle_o = (1 - 2/\phi)\lambda_z^2 + 2/\phi$$

Substituting eq 25 into eq 16 then leads to the elastic free energy of the phantom network as

$$\Delta A_{el} = \frac{\nu kT}{2}(1 - 2/\phi)(\lambda_x^2 + \lambda_y^2 + \lambda_z^2 - 3)$$
$$= \frac{\xi kT}{2}(\lambda_x^2 + \lambda_y^2 + \lambda_z^2 - 3) \qquad (26)$$

The last of eq 26 is obtained by substituting for the cycle rank from eq 1.

8. STRESS-STRAIN RELATIONS FOR THE PHANTOM AND AFFINE NETWORK MODELS IN SIMPLE TENSION

Equations 18 and 26 may be written in compact form as

$$\Delta A_{el} = \mathcal{F}kT(\lambda_x^2 + \lambda_y^2 + \lambda_z^2 - 3) \qquad (27)$$

where $\mathcal{F}$ equates to $\nu/2$ for the affine network model and to $\xi/2$ for the phantom network model. For uniaxial tension, eq 27 takes the following form

$$\Delta A_{el} = \mathcal{F}kT(v_{20}/v_2)^{2/3}(\alpha^2 + 2\alpha^{-1} - 3) \qquad (28)$$

The relationship of the force f along the x direction to ΔA_{el} is given (1) by thermodynamics as

$$f = \partial\Delta A_{el}/\partial L$$
$$= \partial\Delta A_{el}/L_{x,i}\partial\alpha \qquad (29)$$

where $L_{x,i}$ denotes the length of the test specimen in the initial state. Using eq 28 in eq 29 leads to

$$f = 2\left(\frac{\mathcal{F}kT}{L_{x,i}}\right)(v_{20}/v_2)^{2/3}(\alpha - \alpha^{-2}) \tag{30}$$

Experimental data is in general given in terms of the reduced force $[f^*]$ defined as

$$[f^*] = f\, v_2^{1/3}/A_d(\alpha - \alpha^{-2}) \tag{31}$$

Here, A_d represents the dry cross-sectional area of the network. Defined in this manner, the reduced force equates to the shear modulus of the dry network for small strains. Substituting eq 30 into 31 leads to

$$[f^*] = 2\left(\frac{\mathcal{F}kT}{V_d}\right)v_{20}^{2/3} \tag{32}$$

According to eq 32, the reduced force is independent of the degree of deformation. In Figure 3, the reduced force for the phantom and affine networks are shown by the lower and the upper horizontal dashed lines respectively. The points denote results on natural rubber networks (20) tested in tension at various degrees of swelling. The reduced force is clearly seen to decrease with swelling. It is also seen that the reduced force becomes, for this network, to be independent of deformation at a v_2 value of 0.24. The dependence of the natural rubber network on swelling as depicted in Figure 3 is in fact a universal phenomenon for amorphous networks in general (3). The reduced force becomes less dependent on strain upon swelling and becomes approximately independent of swelling at high degrees of dilution. The value of the reduced force at such high dilution equates to that of the phantom network as shown by the lowermost dashed line in Figure 3. The validity of this statement have now been shown by several experimental data on model networks of well known structure (3). At this limit, the reduced force is obtained from eq 32 with $\mathcal{F} = \xi/2$ as

$$[f^*] = (\xi kT/V_d)v_{20}^{2/3} \tag{33}$$

Determination of $[f^*]$ from experimental data thus establishes the value of the cycle rank ξ for the network, which is also related to cross-link density μ/V_d and to molecular weight M_c of chains between cross-links as (3)

$$\begin{aligned}[f^*] &= (\phi/2 - 1)\,\frac{\mu}{V_d}\,kT\,v_{20}^{2/3} \\ &= (1 - 2/\phi)\,\frac{\rho RT}{M_c}\end{aligned} \tag{34}$$

Here, ρ is the density of the bulk polymer and R is the gas constant.

The above discussion illustrates the determination of cross-link density from mechanical data on highly swollen networks. Equations 34

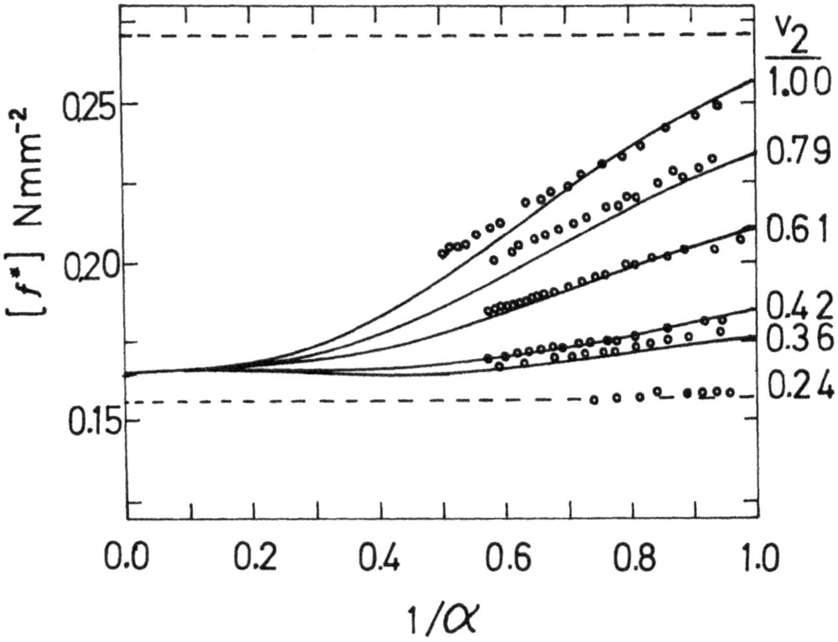

Figure 3. Results of simple tension experiments on natural rubber, from Reference 20. The ordinate represents the reduced force and the abscissa is the reciprocal extension, α^{-1}. The points represent experimental data. Curves are obtained by the constrained junction theory (From Reference 13). Data points are obtained at various degrees of swelling v_2 shown on the right ordinate for each curve. The upper and lower dashed horizontal lines represent the affine and phantom network behavior. The value of the κ parameter used in obtaining the theoretical curves was 8.

are valid for perfect networks only. Their extension to imperfect networks is discussed in some detail in Reference 3.

The curves indicated by solid lines in Figure 3 are obtained by the constrained junction theory (11-13) described in more detail in the following section.

9. THE CONSTRAINED JUNCTION THEORY OF NETWORKS

Experimental data shown in Figure 3 for the dry state and also for smaller degrees of swelling indicate significant departures from the phantom network behavior shown by the lowest horizontal line. One universal nature of these curves is that their extrapolation to $\alpha^{-1} = 0$ leads, approximately, to the phantom network modulus. Thus, performing stress-strain experiments in the dry state and extrapolating [f*] values to $\alpha^{-1} = 0$ by a straight line is another way of determining the degree of cross-linking. This technique corresponds to determination of cross-link density by the use of the Mooney-Rivlin plot which is discussed in large detail in Reference 21. The relationship of the small-strain modulus to network constitution is not understood well, however. We may conclude for the time being that the determination of cross-link density by measuring the phantom network modulus described above is the only established technique.

The theory based on the constrained junction model of networks by Flory is helpful in describing the behavior of networks in the dry state at all levels of deformation. According to this model, the junctions in a real network do not fluctuate freely as they do in the phantom network because of the effect of entanglements present in the real network. The strength of constraints limiting the fluctuations of junctions in real networks is described by a parameter, designated as κ by Flory (11). If there are no constraints, $\kappa = 0$ and the phantom network is obtained. The value of κ increases with increasing degree of constraints. At the other extreme, if the constraints are infinitely strong, the fluctuations of junctions are totally supressed and κ equates to infinity. In this limit the junctions are rigidly embedded in the network, corresponding to the affine network model. Calculations of the reduced force based on this model agree well with experimental data as shown in Figure 3. The constrained junction model thus describes the source of the deviations of real network behavior from that of the phantom networks model. The solid curves that pass through experimental points in Figure 3 are obtained by the constrained junction theory. The value of the κ parameter is chosen as 8 for best fit to all sets of data points at different swelling degrees (A second less important parameter of the theory, ζ, is chosen as 0.12 in the curve fitting. This parameter is not essential for the theory but is introduced to obtaind quantitative agreement with experimental data. The reader is referred to reference 12 and 13 for a detailed discussion of the complete theory).

The constrained junction theory thus completes the gap between the affine and the phantom network formulations. Inasmuch as real networks exhibit properties between affine and phantom networks, the constrained junction model seems to be a satisfactory description.

10. DETERMINATION OF THE DEGREE OF CROSS-LINKING BY EQUILIBRIUM SWELLING MEASUREMENTS

When a network is exposed to the continuous action of a solvent, it swells until equilibrium is reached between the swelling action of the solvent and the elasticity of the network. The volume fraction of polymer in the network swollen to equilibrium is designated by v_{2m}. The phenomenon of swelling of networks immersed in solvents is discussed in the book of Flory (1). However, the formulation there is based on the affine network model. The shortcomings of the affine network model have since been well established (3). The network behaves very close to the phantom network, especially at the high degrees of dilution obtained in the swelling experiment. The formulation based on the affine network model has therefore to be modified to that of the phantom model. The condition that determines the equilibrium degree of swelling of the network is that the chemical potential of the solvent in the network should equate to that in the surrounding pure solvent. This condition leads to the following expression for the phantom network:

$$\Delta\mu_1/RT = \ln(1 - v_{2m}) + \chi v_{2m}^2 + v_{2m} + (V_1/RT)(\xi kT/V_0)(v_{2m}/v_{20})^{1/3} = 0 \qquad (35)$$

Here, $\Delta\mu_1$ is the solvent chemical potential, χ is the Flory-Huggings interaction parameter, V_1 is the molar volume of solvent and V_0 is the volume of network in the state of cross-linking. Measurement of v_{2m} in an equilibrium swelling experiment for known values of χ, V_1, T, and v_{20} thus leads to the evaluation of the cycle rank density ξ/V_0 from eq 35. This value of ξ/V_0 may then be used to estimate the degree of cross-linking of the network. Values of cross-link density obtained from swelling data based on the phantom network model are in general less than those obtained from the affine model. Values of ξ/V_0 obtained from equilibrium swelling measurements are in general in good agreement with those obtained by mechanical experiments discussed in the preceding sections.

11. NUMERICAL EXAMPLE OF CROSS-LINK DENSITY CALCULATION

In this section, a numerical example for determining the cross-link density from stress-strain-swelling data will be worked out using the data given in Figure 3 for natural rubber.

The lower horizontal dashed line, representing the phantom network behavior, has an $[f^*]$ intercept of 0.155 N mm^{-2}. The natural rubber networks of this example were cross-linked in the dry state, and had a functionality of 4. Thus the $[f^*]$ intercept may be related to μ and M_c by the expressions

$$[f^*] = 0.155 \text{ N mm}^{-2}$$

$$= (\mu/V_d)kT \quad (36)$$

$$= \rho RT/2M_c$$

Taking k as $1.381\text{x}10^{-20}$ N mmK^{-1}and T as 300 K, the cross-link density may be obtained from the first and second lines of eq 36 as

$$\mu/V_d = 1.12 \times 10^{19} \text{ cross-links/mm}^3$$

$$= 6.31 \times 10^{-8} \text{ mols of chains/mm}^{-3}$$

Taking $\rho = 0.97$ g cm^{-3}, the average molecular weight M_c between cross-links follows from the first and third lines of eq 36 as

$$M_c = 7800 \text{ g mol}^{-1}$$

The values of μ/V_d and M_c would also follow by extrapolating the data points for $[f^*]$ for the dry network to $\alpha^{-1} = 0$ by a straight line. This operation leads to an $[f^*]$ intercept of 0.15 N mm^{-2}. This value is only 3% less than that obtained from the highly swollen data points.

The value of M_c may also be obtained by equilibrium swelling measurements, by solving eq 35 for M_c as

$$M_c = -\frac{\rho(1 - 2/\phi)V_1 v_{2c}{}^{2/3} v_{2m}{}^{1/3}}{\ln(1 - v_{2m}) + \chi v_{2m}^2 + v_{2m}} \quad (37)$$

Assuming that the solvent in which the network was swollen to equilibrium is benzene with $V_1 = 88$ cm^{3}mol^{-1} and $\chi = 0.44$, a value of $v_{2m} = 0.157$ would lead to $M_c = 7800$ g mol^{-1}.

12. DEVIATIONS FROM IDEAL NETWORK BEHAVIOR

The network model presented in this brief review is based on Gaussian chains. The value of the phantom modulus obtained as $\xi kT/V_o$ thus reflects the properties of networks comprised of such chains. In technological applications, however, behavior of networks deviates from the ideal phantom like behavior in many respects. The most important contribution to modulus in actual networks is due to the presence of fillers such as carbon black, silica, etc. Presently there is no study that relates the amount of fillers to deviations from phantom modulus. Presence of crystallites is a second source of contribution to the modulus. Performing the readings before reaching equilibrium conditions also leads to values of moduli higher than those obtained from equilibrium measurements.

Measurements performed at temperatures close to the glass transition temperature of the network lead to significantly higher moduli values if performed relatively rapidly. Finally, the presence of very short chains in the network leads to a large modulus due to the effect of finite chain extensibility. This effect is not taken into consideration neither in the phantom nor the constrained junction models.

The magnitude of deviations from phantom-like behavior cited in the above paragraph are often very important and may be several times that of the phantom network itself. For this reason, the constitution and the state of the networks should be carefully specified as to the presence of fillers, crystallites, short chains, etc., and care should be exercised in performing the experiments at equilibrium conditions.

13. CONCLUSIONS AND DISCUSSION

The present study emphasizes that the behavior of networks in the highly swollen state resembles that of a phantom network model. The latter model is well understood. Identifying the highly swollen network by the phantom network thus establishes the degree of cross-linking in a straightforward manner. Alternatively, experiments performed on dry networks require extrapolation of data to high extension ratios, where the mechanical modulus becomes close to that of the phantom network model.

The small strain modulus of dry networks does not relate to the degree of cross-linking and should not be adopted for a measure of cross-link density.

Equilibrium swelling measurements lead to cross-link values close to those obtained by mechanical data. However, care should be taken to use eq 35 for this purpose. This equation is based on the phantom network model rather than the incorrect affine network model of previous treatments.

Resemblance of real network behavior to that of the phantom network model in the highly swollen state is established also by neutron scattering experiments. According to these experiments, the directly observed chain dimensions in the swollen deformed network transform close to that calculated from the phantom network model (22). This important observation makes the identification of the swollen network behavior with that of the phantom network more plausible.

REFERENCES

1. Flory, P.J. *"Principles of Polymer Chemistry"*, Cornell University Press, Ithaca, New York, 1953.

2. Treloar, L.R.G. *"The Physics of Rubber Elasticity"*, 3rd ed., Clarendon Press, Oxford, 1975.

3. Mark, J.E. and Erman, B. *"Rubberlike Elasticity: A Molecular Primer"* Wiley, New York, 1988.

4. Wall, F.T. and Flory, P.J. *J. Chem. Phys.*, ***19***, 1435 (1951).

5. James, H.M. *J. Chem. Phys.*, ***15***, 651 (1947).

6. James, H.M. and Guth, E. *J. Chem. Phys.*, ***15***, 669 (1947).

7. Ronca, G. and Allegra, G. *J. Chem. Phys.*, ***63***, 4990 (1975).

8. Deam, R.T. and Edwards, S.F. *Philos. Trans. R. Soc. London*, ***280***, 317 (1976).

9. Ball, R.C., Doi, M., Edwards, S.F. and Warner, M. *Polymer*, ***22***, 1010 (1981).

10. Edwards, S.F. and Vilgis, T. *Polymer*, ***27***, 483 (1986).

11. Flory, P.J. *J. Chem. Phys.*,***66***, 5720 (1977).

12. Flory, P.J. and Erman, B. *Macromolecules*, ***15***, 800 (1982).

13. Erman, B. and Flory, P.J. *Macromolecules*, ***15***, 806 (1982).

14. Eichinger, B.E. *Annu. Rev. Phys. Chem.*, ***34***, 359 (1983).

15. Flory, P.J. *Proc. R. Soc. Lond. A.*, ***351***, 351 (1976).

16. Scanlan, J. *J. Polym. Sci.*, ***43***, 501 (1960).

17. Case, L.C. *J. Polym. Sci.*, ***45***, 397 (1960).

18. Flory, P.J. *Macromolecules*, ***15***, 99 (1982).

19. Flory, P.J. *"Statistical Mechanics of Chain Molecules"*, Wiley-Interscience, New York, 1969.

20. Allen, G., Kirkham, M.J., Padget, J., and Price, C. *Trans. Faraday Soc.*, ***67***, 1278 (1971).

21. Mark, J.E. *Rubber Chem. Technol.*, ***48***, 495 (1975).

22. Bastide, J., Duplessix, R., Picot, C., and Candau, S. *Macromolecules*, ***17***, 83 (1984).

FORMATION AND PROPERTIES OF THERMOSETTING AND HIGH T_g SYSTEMS

John K. Gillham
Polymer Materials Program
Department of Chemical Engineering
Princeton University
Princeton, New Jersey 08544 USA

ABSTRACT. A review of research in the author's laboratory on the conversion of reactive liquids to amorphous polymeric glasses is presented. Interrelationships between reactants, reaction conditions, and subsequent material properties of thermosetting and high T_g polymers are discussed from the point of view of a generalized time-temperature-transformation (TTT) diagram. The TBA torsion pendulum is discussed as a technique for characterizing such polymeric systems.

INTRODUCTION

This review of research in the author's laboratory, which is set into a general context, pertains principally to the formation and properties of high T_g polymeric glasses which are made by the transformation of liquid to amorphous solid by chemical reaction. The area is of particular importance in the making of composites, encapsulants, coatings, and adhesives by thermosetting "cure" reactions in which multifunctional molecules of low molecular weight are converted into network macromolecules. However, an attempt has been made to produce a generalized model for the formation and properties of thermosetting systems which also incorporates linear macromolecular systems. An example of a network-forming system is the cure of neat multifunctional epoxy resin; an example of a linear-forming system is the polymerization of neat difunctional vinyl monomer. An example of a thermosetting system which results in a linear polymer is the formation of semi-ladder linear polyimide by intramolecular reaction of linear polyamic acid. In each case, polymerization below the glass transition temperature of the fully reacted material results in a glass at the temperature of reaction which is the consequence of the glass transition temperature rising to the temperature of reaction. The review will therefore emphasize thermosetting systems with some reference to linear systems. The simplest model for chemical setting assumes a single reaction mechanism and no phase separation.

O. Güven (ed.), Crosslinking and Scission in Polymers, 171–198.

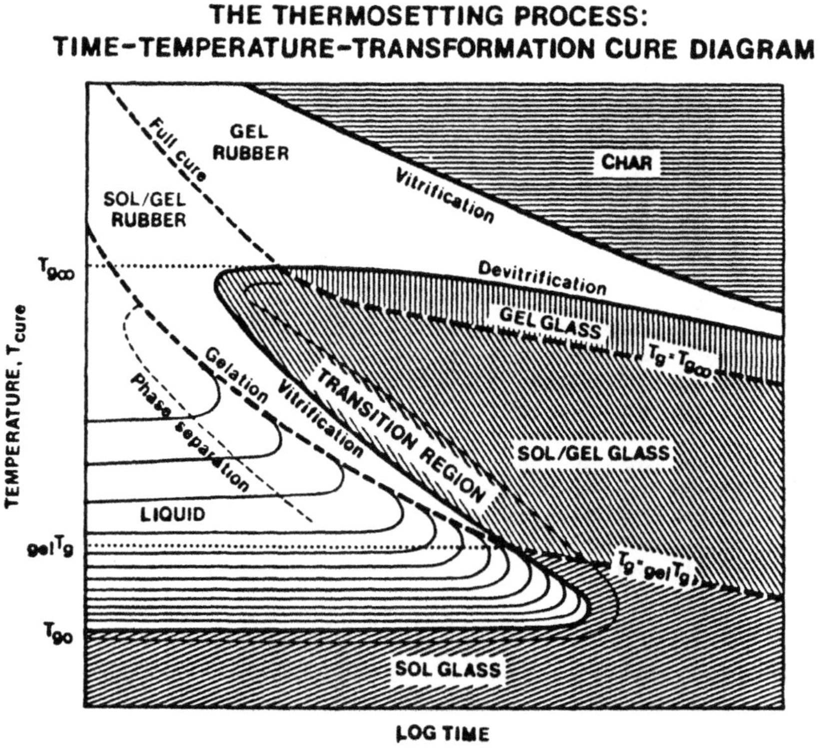

Figure 1. Schematic Time-temperature-transformation (TTT) isothermal cure diagram for a thermosetting system, showing three critical temperatures, i.e., $T_{g\infty}$, ${}_{gel}T_g$, T_{go}, and the distinct states of matter, i.e., liquid, sol/gel rubber, gel rubber (elastomer), gelled glass, ungelled (or sol) glass, and char. The full-cure line, i.e., $T_g = T_{g\infty}$, divides the gelled glass region into two parts: sol/gel glass and fully cured gel glass. Phase separation occurs prior to gelation. Successive isoviscous contours shown in the liquid region differ by a factor of ten. The vitrification process below ${}_{gel}T_g$ has been constructed to be an isoviscous one. The transition region approximates the half width of the glass transition.

THE TIME-TEMPERATURE-TRANSFORMATION (TTT) ISOTHERMAL CURE DIAGRAM

A time-temperature-transformation (TTT) isothermal cure diagram (Fig. 1) may be used to provide an intellectual framework for understanding and comparing the cure and physical properties of thermosetting systems (1,2). The main features of such a diagram can be obtained by measuring the times to events that occur during isothermal cure at different temperatures, T_{cure}. These events include the onset of phase separation, gelation, vitrification, full cure, and devitrification. Phase separation may occur, for example, by precipitation of rubber during cure of rubber-modified formulations, by the formation of monomer-insoluble oligomeric species and by the formation of gel particles (and, in crystallizable systems, by crystallization). Molecular gelation corresponds to the incipient formation of molecules of infinite molecular weight, which gives rise to macroscopic gelation in the macroscopic fluid. According to Flory's theory of gelation (3), for the simplest system, molecular gelation occurs at a definite conversion which is independent of temperature. After molecular gelation the material consists of normally miscible sol (finite molecular weight) and gel (infinite molecular weight) fractions, the ratio of the former to the latter decreasing with conversion. Molecular gelation does not correspond precisely to macroscopic gelation (see later). Vitrification occurs when the glass-transition temperature, T_g, rises to the temperature of cure. The material is liquid or rubbery when $T_{cure} > T_g$; it is glassy when $T_{cure} < T_g$. Devitrification occurs when the glass transition temperature decreases through the cure temperature, as in some types of thermal degradation. The diagram displays the distinct states encountered on cure due to chemical reactions. These states include liquid, sol/gel rubber, gel rubber (elastomer), ungelled (sol) glass, gelled glass, and char. The gelled glass region in the TTT cure diagram is divided into two parts by the full-cure line; in the absence of degradation (Fig. 1, devitrification and char), the top and lower parts can be designated fully cured gel glass and undercured sol/gel glass regions, respectively. The technological terms, A-, B- and C-stage resins correspond to sol glass, sol/gel glass, and fully cured gel glass, respectively. The illustration also displays the critical temperatures $T_{g\infty}$, ${}_{gel}T_g$, and T_{go}, which are the glass-transition temperature of the fully cured system, the temperature at which molecular gelation and vitrification occur simultaneously, and the glass-transition temperature of the reactants, respectively. The temperature at which macroscopic gelation and vitrification occur simultaneously is designated ${}_{gel}T_g'$.

Much of the behavior of thermosetting materials can be understood in terms of the TTT cure diagram through the influence of gelation, vitrification, and devitrification upon properties. Gelation retards macroscopic flow, and retards growth of a dispersed phase (e.g., as in rubber-modified systems). Vitrification retards chemical conversion. Devitrification, due to thermal degradation, marks the lifetime for the material to support a substantial load.

The isothermal TTT cure diagram for the curing of linear

elastomeric systems is of more limited use than for thermosetting systems because, in practice, only the region above $T_{g\infty}$ is relevant for the former. Gelation in the vulcanization of rubbers occurs at low conversions in comparison with typical thermosetting systems.

The ungelled glassy state is the basis of commercial molding materials since, upon heating, the ungelled (sol) material flows before gelling through further reaction. Formulations can be processed as solids (e.g., molding compositions) when T_{go} > ambient temperature; they can be processed as liquids (e.g., as casting fluids) when T_{go} < ambient temperature.

The glass transition temperature of the material at the composition corresponding to molecular gelation is ${}_{gel}T_g$, since molecular gelation occurs as the material vitrifies when the temperature of cure is ${}_{gel}T_g$ (4). The molecular gelation curve (Fig. 1) therefore corresponds to $T_g = {}_{gel}T_g$. Temperature ${}_{gel}T_g$ is critical in determining the upper temperature for storing reactive materials in order to avoid gelation (which relates to "pot life"). However, cure below temperature ${}_{gel}T_g$ eventually leads to gelation.

The morphology developed in two-phase systems, for example, those in which rubber-rich domains nucleate and grow as a dispersed phase in a curing rubber-modified thermoset, depends on the temperature of cure. The reaction temperature determines the competition between thermodynamic and kinetic (i.e., transport) factors which affects the amount, the composition, and the distribution of dimensions of the dispersed phase. As an example, reaction at intermediate temperatures can result in a maximum in the amount of precipitated phase which may (or may not) accompany a minimum in the time to the onset of phase separation (as shown in Fig. 1). For optimum mechanical properties, a two-phase system is cured first at one temperature to provide a particular morphology, and subsequently at a higher temperature to complete the reactions of the matrix (5,6). The $T_{g\infty}$ of the matrix will be determined by the extent of phase separation. A matrix parameter, the critical interparticle distance, has been reported (7) which determines the brittle to ductile transition in rubber-modified thermoplastics. For a given volume fraction of a dispersed low-modulus phase the critical interparticle distance is determined by the particle size. This concept should be applicable to rubber-modified thermosets in which cure under different conditions affects both the amount of phase separation and the particle size distribution.

In composite systems, shrinkage stresses due to volume contraction of the resin on isothermal cure begin to develop with adhesion of the curing resin to a rigid inclusion or substrate. This occurs after gelation above ${}_{gel}T_g$ and before vitrification below ${}_{gel}T_g$. The tensile stresses in the resin and the corresponding compressive stresses on an inclusion and substrate affect composite behavior. One consequence is fiber-buckling in resin/fiber composites. A related consequence of the shrinkage due to cure, and of the different coefficients of expansion and contraction of the constituents in brittle resin/fiber composites, is the formation of spiral and helical cracks in the resin around isolated filaments and yarns (8,9). Their large surface areas per unit volume of matrix may contribute to the toughening of fiber/resin composites.

Prolonged isothermal cure at temperature T_{cure} below $T_{g\infty}$ would lead to $T_g = T_{cure}$ if the reactions were quenched by the process of vitrification. In practice, T_g is higher than T_{cure} because it can increase during the heating scan employed for measurement subsequent to cure. Although vitrification has been defined to occur when $T_g = T_{cure}$, the glass transition temperature, T_g, as usually measured, does not correspond to the glassy state, but rather to a state approximately halfway between the rubbery and glassy states; therefore reactions at T_{cure} continue beyond the assigned time to vitrification, which also results in $T_g > T_{cure}$. Furthermore, the extents to which reactions proceed in the glassy state depend on the influence of the glassy state on the reaction kinetics. However, even the intramolecular reactions involved in the imidization of polyamic acids to polyimides are restricted by the vitrification process, leading again to T_g being controlled by the temperature and the time of cure (10). In practice, for epoxies and polyimides, T_g is greater than T_{cure} by about 30-50°C after "normal" isothermal cure below $T_{g\infty}$ (10-13). This corresponds approximately to the half-width of the glass transition temperature region since T_g (as measured) increases through the isothermal temperature T_{cure} to about T_{cure} + 30 to 50°C after which the reaction rate is controlled by the low physical relaxation rates of the glassy state. The post vitrification glass transition region, which is indicated very qualitatively in Figure 1, varies since the width of the glass transition depends on the value of T_g and since the time scale depends on the reaction rate and on the concentration of reactants prior to vitrification.

Correlations between macroscopic behavior and molecular structure of the reactants and of the product are most clearly defined in fully cured materials. Full cure is attained most readily by reaction above $T_{g\infty}$, and more slowly by curing below $T_{g\infty}$ to the full-cure line of the TTT cure diagram (12). The full cure line corresponds to $T_g = T_{g\infty}$.

In practice full cure is, in general, not a unique state because the state depends upon the time-temperature reaction path. In commonly used systems this is a consequence of competing chemical reactions with different activation energies. Furthermore, the time-temperature path of cooling after cure affects, for example, density and material behavior at room temperature.

At high temperatures, non-curing chemical reactions result in degradation. Thermal degradation can result in devitrification as the glass-transition temperature decreases through the isothermal temperature due to a reduction in cross-linking or formation of plasticizing material. Degradation can also result in vitrification, e.g., char formation (Fig. 1), as the glass-transition temperature increases through the isothermal temperature because of an increase in cross-linking or volatilization of low molecular weight plasticizing materials (13). Cure and thermal degradation reactions compete in high T_g systems. There is a need to obtain high temperature polymers from low temperature processing.

The limiting viscosity in the fluid state is controlled by molecular gelation above ${}_{gel}T_g$, and by vitrification below ${}_{gel}T_g$. At gelation, the weight-average molecular weight and zero shear-rate viscosity become infinite, although the number-average molecular weight is generally low. Viscosity prior to vitrification below ${}_{gel}T_g$ is described by the Williams-Landel-Ferry (WLF) equation (14,15).

The time to reach a specified viscosity (Fig. 1) is often used as a practical method for measuring gelation times. Although this macroscopic isoviscosity approach is inconsistent with the molecular isoconversion theory of gelation, above temperature ${}_{gel}T_g$ the apparent activation energies obtained from the temperature dependence of the time to reach a specified viscosity converge to the true activation energy for the chemical reactions leading to molecular gelation with increase of the specified viscosity (1).

The time to molecular gelation can be computed from the reaction kinetics and the conversion at molecular gelation, which is constant according to Flory's theory of gelation. The time to vitrification can be computed (1,15-17) from the reaction kinetics and the conversion at vitrification (which increases with T_{cure}). Since vitrification occurs when the glass-transition temperature reaches the temperature of cure, computation of the time to vitrify requires knowledge of the relationship between T_g and conversion. Figure 2 shows that T_g increases with conversion at an increasing rate. In the absence of diffusion control, the simplest kinetic equation describing the reaction is

$$dX/dt = [A\exp(-E/RT)]f(X) \qquad [1]$$

where X is the fractional extent of reaction, E the activation energy, and the other characters have their usual significance. The times to molecular gelation and to vitrification can be computed versus temperature using this equation when X_{gel} (for gelation), the relationship between X and T_g (for vitrification), and the reaction kinetics are known.

The S-shaped vitrification curve obtained experimentally in the absence of thermal degradation has been matched by computation for one epoxy system from temperature T_{go} to temperature $T_{g\infty}$ (15). Comparison of measured and computed times to vitrification for epoxy systems indicates that reactions are diffusion controlled only after vitrification (15).

The vitrification curve is generally S-shaped (17). At temperatures immediately above T_{go}, the time to vitrify passes through a maximum because of the opposing influences of the temperature dependences of the viscosity and the reaction rate constant (1). Immediately below $T_{g\infty}$, the time to vitrify passes through a minimum (18) because of the opposing influences of the temperature dependence of the reaction rate constant and the decreasing concentration of reactive sites at vitrification as $T_{g\infty}$ is approached. Knowledge of the minimum time and the corresponding temperature is useful in molding technology when cured specimens must be ejected in the glassy state.

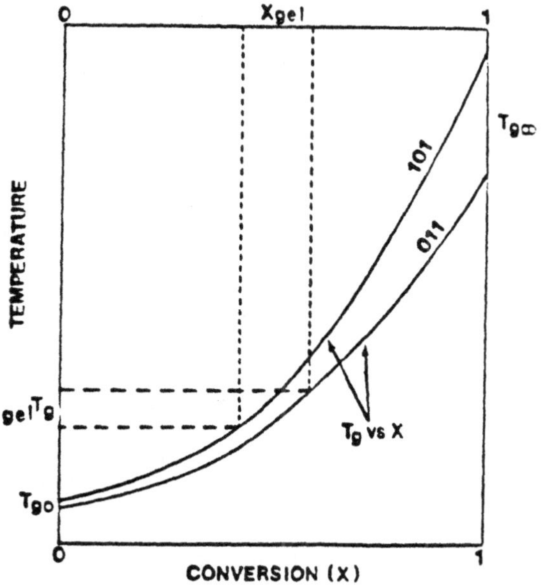

Figure 2. T_g vs. conversion at vitrification for reactants differing in functionality (101 > 011). The conversions at molecular gelation are also included. The diagram can be used to demonstrate the effect of increasing functionality of the reactants on molecular gelation, vitrification, and the temperatures $_{gel}T_g$, T_{go}, and $T_{g\infty}$.

Cure of finite specimens at temperatures below that for the minimum time for vitrification (Fig. 1) can lead to the hotter inside vitrifying before the outside when the reaction is exothermic. Conversely, cure at higher temperatures can lead to the outside vitrifying before the inside: in this case internal stresses develop as the inside contracts relative to the vitrified outside due to the volume contraction of polymerization. The situation can be similar to the latter case when external heating causes the hotter outside to vitrify before the cooler inside. Similarly, cooling cured material from above $T_{g\infty}$ will lead to the outside vitrifying before the inside; internal stresses in this case will be minimized by cooling very slowly (i.e., annealing). These stresses, which are due to non-isothermal conditions in neat systems, are supplemented in composites by the differential shrinkage stresses which were referred to earlier in the article.

The conversion at vitrification can be computed in principle by relating the glass-transition temperature to contributions from the molecular weight and the cross-linking density, both of which vary with conversion (17). For polymerization prior to gelation (and for linear polymerization), the computation is simplified by the absence of cross-linking.

The fractional extent of reaction at vitrification and the time to vitrify, like molecular gelation, decrease with increasing functionality of the reactants (13). The effect of increasing functionality on molecular gelation, vitrification, and the temperatures $_{gel}T_g$, T_{go}, and $T_{g\infty}$ can be understood by considering the X-vs-T_g and the X_{gel} relationships, such as those in Figure 2. For example, the figure shows how temperature $_{gel}T_g$ for the material of lower functionality can be higher than that of higher functionality.

ANOMALOUS PROPERTIES VERSUS EXTENT OF CURE

Increasing reaction time at an elevated temperature leads to increasing conversion, T_g, average molecular weight, and cross-linking density. The modulus and density at the curing temperature also increase with increased conversion. However, the room temperature [RT] modulus and RT density can decrease, whereas the equilibrium absorption of water at RT can increase with increasing extent of cure. The anomalous behavior of density can result, for example, in a net expansion at RT when a reactive material which has been set at RT is post-cured at elevated temperatures and cooled to RT. A common basis for these interrelated phenomena is the increasing free volume at room temperature with increasing extent of cure (19,20).

The room temperature density (ρ_{RT}) actually passes through a maximum value with increasing conversion (T_g) in the vicinity of gelation. Figure 3 includes a "master" curve of ρ_{RT} versus T_g - RT (20) for specimens which had cooled slowly to RT. (It is difficult experimentally to obtain a complete ρ_{RT} versus conversion curve from one cure temperature for "slow-cooled" specimens, since the reaction rate is high at high temperatures of cure, and vitrification limits the extent

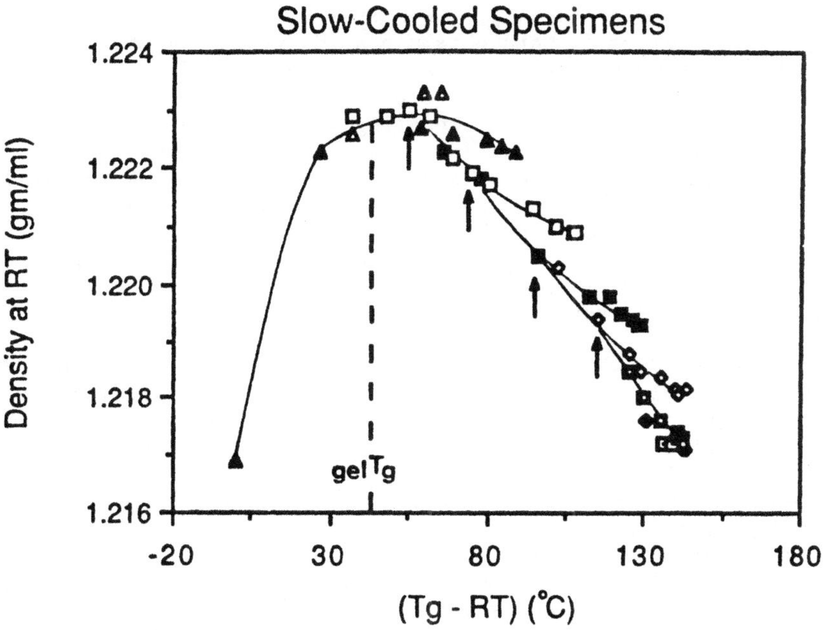

Figure 3. RT density of slow-cooled specimens vs. (T_g - RT) for different isothermal cure temperatures (20): (△) 66.7°C; (▲) 80°C; (□) 100°C; (■) 120°C; (◇) 140°C; (▣) 165°C; (◆) 180°C, and (▣) 200°C. Arrows indicate vitrification where T_g = T_{cure}. Temperature ${}_{gel}T_g$ (dashed line) is used as a demarcation between ungelled and gelled specimens.

of reaction below temperature $T_{g\infty}$. A master curve was therefore constructed from ρ_{RT} vs T_g data obtained at different isothermal temperatures by assuming that T_g is a direct measure of conversion in the absence of vitrification during cure.) Perturbations from the master curve are due to physical annealing after isothermal vitrification which (after gelation) prevents the density from decreasing to the level of the master curve with increasing conversion. Vitrification nullifies the unique relationship between ρ_{RT} and conversion for a given cooling rate. The same value of ρ_{RT} can then arise from different combinations of conversion and extent of physical annealing (aging); for example, a particular value of ρ_{RT} can be obtained at lower temperatures with less conversion but more physical annealing.

The limiting ρ_{RT} which corresponds to the highest T_g attained after long times beyond vitrification at each temperature of cure depends on T_{cure} and varies inversely with T_{cure} (when $T_{cure} < T_{g\infty}$) (20). When T_g is high, the limiting ρ_{RT} will be stable at room temperature (if the cooling rate from T_{cure} to RT is low).

Although the maximum difference in ρ_{RT} which can be obtained by curing to different extents is only about 0.5 percent (Fig. 3), this corresponds to differences of about 20 percent in the free volume of the glass (20). It is not surprising therefore to observe corresponding changes of 20 percent in modulus and water absorption at room temperature. Furthermore, the stresses at polymer/substrate interfaces, which determine failure in composites, depend directly on the modulus of the polymer as well as on the differential dimensional changes of the components (20). These concepts are relevant to the design of plastic materials with the dimensional stability and stength needed for the replacement of metals by plastics.

One factor contributing to the maximum in ρ_{RT} versus conversion is summarized as follows (20). The value of ρ_{RT} depends on the contraction due to chemical reaction and the subsequent contraction due to cooling from T_{cure} to RT. The latter involves contraction in the rubbery state (from T_{cure} to T_g) and contraction in the glassy state (from T_g to RT). With higher values of T_g there is less contraction on cooling since the coefficient of contraction of the rubbery state is higher than that of the glassy state. Since T_g rises nonlinearly with increasing conversion (Fig. 2), higher conversions can give rise to even less contraction on cooling from T_{cure} than would be anticipated from linear behavior. The net sum of cure shrinkage and temperature shrinkage can therefore change from positive to negative with increasing conversion. The maximum in ρ_{RT} is associated with gelation (and therefore, as in Fig. 3, with the glass transition temperature ${}_{gel}T_g$ at the composition corresponding to gelation) since the nonlinearity in T_g versus conversion relationship is amplified by the increase of crosslink density which occurs progressively after gelation with increasing conversion.

Another factor contributing to ρ_{RT} decreasing with increasing conversion after gelation is the increasing relaxation times in the glass transition region with increasing T_g; cooling at a constant rate from T_{cure} results in RT glasses which are progressively further from equilibrium. The higher the glass transition temperature, the further from equilibrium a glass is at RT.

THE CONTINUOUS HEATING TIME-TEMPERATURE-TRANSFORMATION (CHT) CURE DIAGRAM

The concept of the TTT cure diagram can be extended to non-isothermal conditions for curing and annealing polymeric materials. For example, a continuous heating time-temperature-transformation (CHT) diagram results from heating a reactive system from (say) 25° ($<T_{go}$) to 300°C ($> T_{g\infty}$) at a series of different rates of increasing temperature (T). Devitrification at the glass transition temperatue of the reactants (T_{go}) will be joined to the vitrification curve due to cure (i.e., T_g = T) which in turn will be joined to a devitrification curve (i.e., T_g = T) to form an envelope (21). This latter devitrification is a consequence of the rising glass transition temperature after vitrification eventually not increasing at the same rate as the rate of rise of temperature.

The CHT diagram is useful in molding technology for defining time-temperature cure paths in which vitrification does not occur, for example, so as to obtain full cure. Conversely vitrification can be an essential part of a cure cycle so as to control reaction rates which could run out of control because of the exothermic nature of a reaction. The polymerization of over one ton of an epoxy resin system to encapsulate a magnetic coil in Princeton University's experimental Tokomak nuclear fusion reactor is accomplished by heating at a very low rate of temperature increase; the reaction rate is controlled in the process by the T_g increasing in concert with the temperature until $T_{g\infty}$, i.e., full cure, is attained. For heating rates below a particular value (for a given system) an unreacted formulation can be heated such that it does not devitrify, the reaction rate then being sufficient to keep the increasing glass transition temperature equal to or greater than the ambient temperature, until the fully reacted system devitrifies through its maximum glass transition temperature. This concept has implications for controlling material properties in which volume changes are relevant.

LINEAR POLYMERIZATION AND THE TTT DIAGRAM

This section attempts, in a general manner, to incorporate linear polymerization of neat systems into the concept of the TTT diagram.

Difunctional liquid monomers can be transformed into glassy linear polymeric materials by reaction below the maximum glass transition temperature of the polymer, $T_{g\infty}$. The time to vitrification curve for the isothermal free radical polymerization of neat styrene versus the isothermal temperature of reaction, T_{rx}, has been computed in the absence of diffusion control (22). The calculation involved using the well established chemical kinetics from zero conversion to the conversion corresponding to $T_g = T_{rx}$. Since the composition at any time consists of monomer and polymer (neglecting initiator and active chain radicals), the proportion of polymer to monomer (i.e., the conversion) was calculated from the glass transition temperatures of the monomer (-

100°C) and polymer (+100°C) and from $T_g = T_{cure}$, using a simple law of mixtures. The computed vitrification curves for the free-radical polymerization of styrene, and also for a linear-forming step-growth polymerization, were S-shaped.

Phase separation can occur in neat linear polymerization as a consequence of insolubility of oligomer or polymer. This has been a particularly important consideration in the synthesis by step-growth polymerization of high T_g linear polymers, which of necessity have relatively inflexible chain structures. Precipitation often effectively removes the growing species from the reaction medium, thereby limiting the molecular weight. In the same type of polymerization the reaction temperature must be such that vitrification also does not occur (i.e. $T_{rx} > T_{g\infty}$). Phase separation can be analogous to vitrification in limiting molecular weight and T_g. (Use of appropriate solvents overcomes the problems of insolubility and vitrification in the synthesis of high T_g rigid molecules: however, these solvents must be subsequently removed, often by ingeneous methods.)

Although chemical gelation does not occur in linear polymerization, long range elasticity can develop in the polymerizing fluid as the consequence of entanglements between polymer molecules. These will occur at lower conversions for chain reactions where high molecular weight polymer forms throughout the reaction, than for step-growth polymerizations where high molecular weight polymer develops only late in polymerization.

The situation is more complex for the conversion of liquid monomers into polymers which can crystallize. Polymerization of neat monomer above $T_{g\infty}$ can lead to solidification because of crystallization. (The maximum melting temperature, $T_{m\infty}$, is greater than the maximum glass transition temperature, $T_{g\infty}$.) Polymerization below $T_{g\infty}$ can lead to solidification as the result of vitrification or crystallization. The melting temperature range, as well as the glass transition temperature are directly related to the temperature of reaction: it is well known that T_m of a linear polymer is directly related to the isothermal temperature of crystallization.

SOLVENT-BASED REACTIVE SYSTEMS AND THE TTT DIAGRAM

An important application of the transformation of reactive liquids to glassy solids involves loss of solvent. Vitrification can occur as the consequence of both solvent loss and chemical reaction (10,23). The following example demonstrates how the methodology of the isothermal TTT cure diagram has been used to characterize a particular solvent-based reactive coating (23). The isothermal TTT diagram is shown as Figure 4.

The particular system is reactive at reasonable rates only above $T_{g\infty}$ ($T_{g\infty} \simeq 147$°C). Cure involves volatilization of both solvent (57.5 percent) and reaction by-products (4-6 percent). Although, in practice, isothermal vitrification is therefore not a consideration, complete characterization of a system would include it, as is shown in Figure 4.

The progress of the changes which occur are summarized by iso-

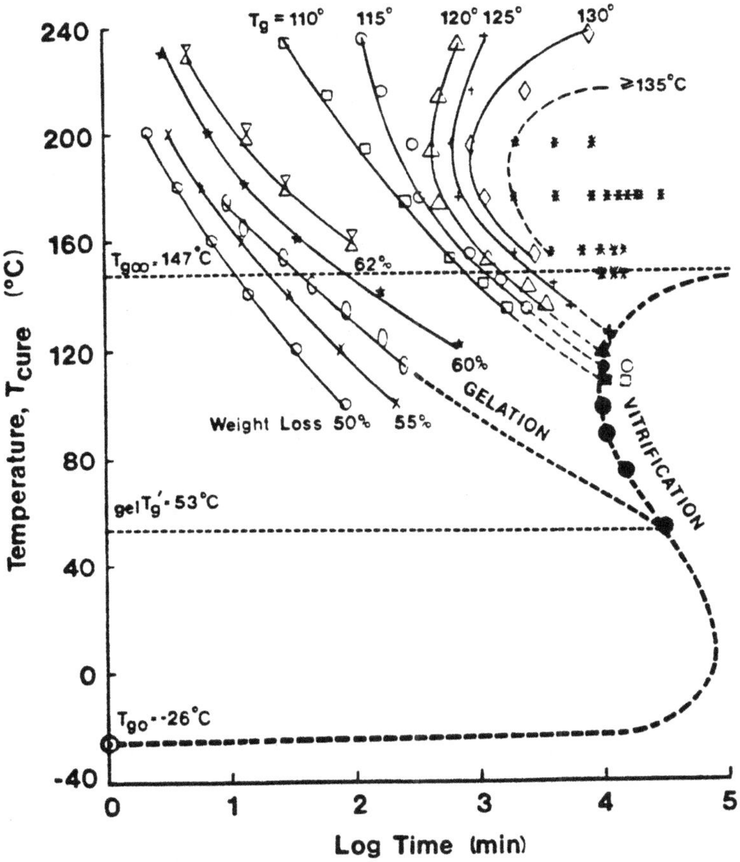

Figure 4. Summary Time-Temperature-Transformation (TTT) Diagram for a Solvent-Based Reactive Coating System. (Glass transition temperatures ⩾ 135°C are bounded as marked.)

weight loss contours (obtained from crossplotting isothermal gravimetric analyses), iso-T_g contours (obtained from crossplotting isothermal changes of T_g, and from analysis of T_{cure}, t_{cure} and T_g data), contours corresponding to macroscopic gelation (obtained directly), and to vitrification. Several data points for vitrification were obtained by direct measurement, including T_{go} which was obtained by direct measurement on the unreacted formulation during a temperature scan. However, the vitrification contour was obtained, for the most part, by extrapolation techniques [a) extrapolating iso-T_g contours to T_{cure} = T_g; b) analyzing T_{cure}, t_{cure} and T_g data; c) extrapolating the macroscopic gelation contour such that the time to macroscopic gelation equals the time to vitrification to give the time to vitrify at temperature ${}_{ge}T_g'$, and the temperature ${}_{gel}T_g'$].

The iso-weight loss contours in the TTT diagram indicate that solvent is essentially lost before gelation occurs. (Retention of solvent beyond gelation is not generally desirable.) Subsequent loss of weight is attributed to the byproducts of the chemical reaction.

The iso-T_g contours would correspond to iso-conversion contours if T_g is a direct measure of conversion (Fig. 2). The conversion times to a fixed T_g would then be expected to decrease with increasing temperature in an Arrhenius fashion (see T_g = 110°C contour). However, at high temperatures of cure, thermal degradation is considered to compete with chemical cure to produce the minima in time to reach iso-T_g levels (e.g., see iso-T_g contour = 130°C). Competition between cure and thermal degradation therefore renders it difficult to define $T_{g\infty}$, the maximum glass transition temperature due to only full cure. The maximum experimental value of T_g obtained isothermally (147°C) was taken to be the operational value of $T_{g\infty}$. Degradation was also held responsible for the maximum attainable value of T_g decreasing with increase of T_{cure} (e.g., see T_g = 135°C and T_g = 130°C contours).

THE GLASS TRANSITION TEMPERATURE, T_g, AS A CHARACTERIZING PARAMETER FOR REACTIVE SYSTEMS

T_g is an important and useful parameter in the TTT and CHT cure diagrams. For example, $T_g = T_{cure}$ for vitrification, devitrification and char formation. The molecular state of the unreacted material has $T_g = T_{go}$, and that of the fully reacted material has $T_g = T_{g\infty}$. Molecular gelation corresponds to a definite conversion for the simplest system, for which $T_g = {}_{gel}T_g$. Macroscopic gelation (corresponding, for example, to an isoviscous state) occurs simultaneously with vitrification at $T_g = T_{cure} = {}_{gel}T_g'$.

The basic parameter governing the state of the setting of a reactive material is the chemical conversion. Therefore more details of the cure process can be visualized by incorporating iso-conversion contours in the TTT and CHT cure diagrams. However, the sensitivity of spectroscopic and calorimetric techniques for measurements of small changes in conversion in complex systems (e.g., epoxies and polyimides), is inadequate, especially at high conversions. Current research in our laboratory is directed to study the feasibility of using T_g as a direct one-to-one measure of conversion, and to employ a simple methodology for constructing iso-T_g contours and also the vitrification contour in the TTT and CHT diagrams from T_g versus time data at different values of T_{cure}, providing that a one-to-one relationship between T_g and conversion exists (23,24).

A derivation of the relationship between the times to reach a particular T_g at different isothermal cure temperatures follows (23,24) assuming a one-to-one relationship between T_g and fractional conversion (X). From the integral form of the rate expression (equation [1]),

$$\int_0^X \frac{dX}{f(X)} = \left[A \exp\left(\frac{-E}{RT}\right)\right] t_X \qquad [2]$$

where t_X is the time to reach conversion X at temperature T. The left hand side of the equation is a function of conversion (X), and thus, also a function of only T_g [$\equiv F(T_g)$]. The time to reach a fixed conversion, t_X, can be replaced by the time to reach the corresponding T_g, t_{Tg}. Thus the above equation can be written as

$$F(T_g) = \left[A \exp\left(\frac{-E}{RT}\right)\right] t_{Tg} \qquad [3]$$

This equation is valid for all cure temperatures for the simplest systems in which the reaction remains kinetically controlled.

Let $t_{Tg,1}$ be the time needed to reach a given T_g at cure temperature T_1 and let $t_{Tg,2}$ be the time needed to reach the same T_g at cure temperature T_2.

$$\therefore \quad \left[A \exp\left(\frac{-E}{RT_1}\right)\right] t_{Tg,1} = \left[A \exp\left(\frac{-E}{RT_2}\right)\right] t_{Tg,2} \qquad [4]$$

and

$$\frac{-E}{RT_1} + \ln(t_{Tg,1}) = \frac{-E}{RT_2} + \ln(t_{Tg,2}) \qquad [5]$$

This is the required relationship for an iso-T_g contour in the isothermal TTT cure diagram. Thus, if the time to reach a particular T_g at one cure temperature is known, then the times to reach the same T_g at different temperatures can be calculated provided that the reaction activation energy is available. It also follows that if the time to reach all T_g values at a single temperature of cure (e.g. at $T_{g\infty}$) are obtained, then iso-T_g contours can be calculated for all values of T_g. The activation energy can be obtained simply from Arrhenius-type plots of the logarithm of the times to reach a definite T_g, in the absence of vitrification, versus reciprocal temperature (K). The vitrification contour is obtained from iso-T_g contours using the relationship $T_g = T_{cure}$.

The glass transition temperature, T_g, is a sensitive and practical parameter for following the cure of reactive thermosetting systems. A wide range of values of T_g is encountered during cure [e.g., 200°C for high T_g epoxies], and it can be measured easily and accurately [e.g. < 0.5°C] throughout the entire range of cure, especially by dynamic mechanical methods. The fact that T_g increases nonlinearly with conversion in crosslinking systems renders it even more sensitive in the later stages of reaction. Sensitivity is required especially when the reaction rate is low, for example at high conversion, and after solidification (vitrification).

T_g can be used to monitor cure in the diffusion-controlled regime beyond vitrification. Since the value of T_g beyond vitrification depends on chemical aging (conversion) and also on physical aging (annealing), the technique for measuring T_g should separate the two contributions. T_g can then be used as a measure of conversion in the glass transition region and in the glassy state to investigate the dependence of the process of vitrification on chemical conversion (24). Similarly the kinetics of physical aging after vitrification can be studied as a function of temperature (and conversion).

TORSION PENDULUM (TP) AND TORSIONAL BRAID ANALYSIS (TBA): TECHNIQUES FOR CHARACTERIZING THE CURE AND PROPERTIES OF THERMOSETTING SYSTEMS

This section of the article deals with the principal technique -- TBA - which has been developed and used principally in the author's laboratories to characterize the cure and properties of polymeric systems. From the point of view of this review, the technique is especially useful in measuring glass transition temperatures, conveniently and accurately, including under conditions where the material is not self-supporting. Developments in instrumentation and computer software are such that the sensitivity of the technique is limited by the control of temperature, which is facilitated by use of a large copper block as the thermal mass.

TP/TBA - MODES

The freely oscillating torsion pendulum can be used in two ways to characterize polymeric systems: the conventional torsion pendulum (TP) mode and the torsional braid analysis (TBA) mode, respectively (1). Geometrically simple specimens (e.g., films & filaments) are used in the TP mode to obtain quantitative values of the elastic and loss moduli. Impregnated braid specimens are used in the TBA mode to obtain transitions of polymers and oligomers, to obtain a relative mechanical assessment of polymers, and to characterize liquid to solid transformations (e.g. cure). It should be noted that, for a TBA specimen in which the ratio of polymer to substrate is constant (as for specimens prepared from a single polymer/solvent system), T_g occurs at a constant frequency (typically 0.7 Hz).

INSTRUMENTATION

A schematic diagram of the automated TBA torsion pendulum system is shown in Figure 5. (The system is available from Plastics Analysis Instruments, Inc., Princeton, NJ 08540.) The upper activation rod, which is clamped to the upper end of the specimen, is held vertically in alignment with a bushing (Fig. 5, not shown) and is rigidly attached at its upper end to a horizontal gear which, when intermittently subjected to an angular step-displacement,intermittently initiates free torsional oscillations in the specimen. The lower pendulum rod, which is clamped to the lower end of the specimen, hangs freely and is magnetically coupled to a polaroid disc (Fig. 5, polarizer) at its lower end which serves as inertial mass and part of the optical transducer.

The pendulum is enclosed in a gas-tight cylindrical chamber (0.5" diameter), the atmosphere of which is closely controlled and monitored. Inert, water-doped, and reactive gases have been used. There are no electronic devices within the specimen chamber. Dry helium, rather than nitrogen, is usually used as an inert atmosphere because of its higher thermal conductivity at cryogenic temperatures. An on-line electronic hygrometer is used to continuously monitor the water vapor content of atmospheres from <20 to 20,000 ppm H_2O. A cylindrical copper block, wound around with cooling coils for liquid nitrogen and with band heaters, surrounds the specimen. Excellent temperature control is achieved by virtue of the large thermal mass of the copper block. A temperature programmer/controller system permits experiments to be performed in isothermal (±0.1°C above 30°C) and dynamic modes from -190 to 400°C; in the dynamic mode the temperature may be increased or decreased linearly at controlled rates. Cooling is achieved by controlling the flow of liquid nitrogen from a pressurized container. Routine temperature scans are made at rates of change of temperature of ±2°C/minute. Measurements have been made as low as 4°K and as high as 700°C in modified apparatus.

A key factor in the instrumentation is the non-drag optical transducer which produces an electrical response that varies linearly

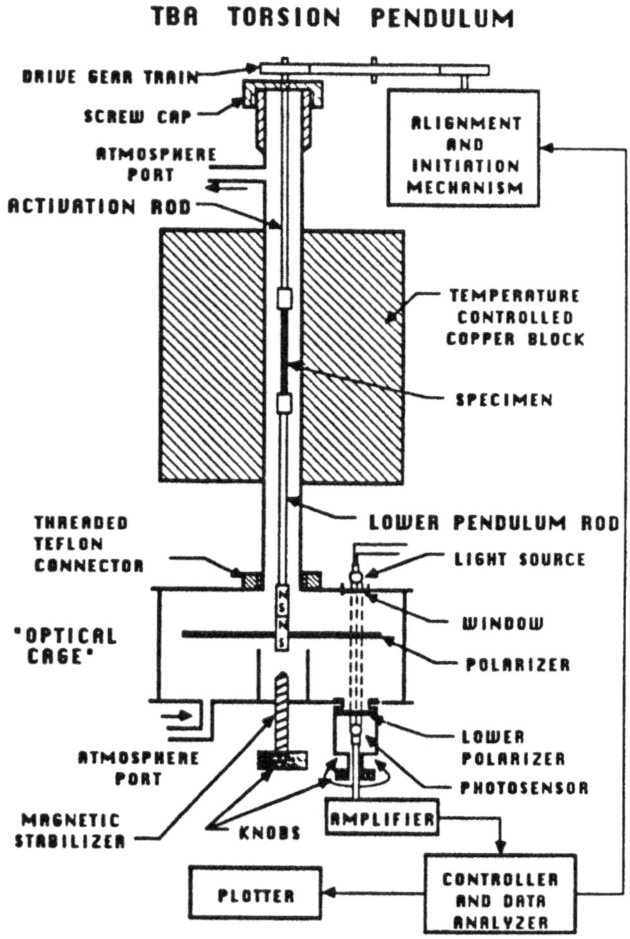

Figure 5. TBA Torsion Pendulum.

with angular displacement. A polarizing disc is used as part of the inertial member of the pendulum, and a stationary second polarizer is positioned in front of a linearly-responding photo-cell (Fig. 5): the transducer is linear when light passes through the aligned pair of polarizers. An analog electrical signal is obtained from a light beam passing through the pair of polarizers.

The clamps on the pendulum allow a variety of specimens to be used: film, fiber, or coating on glass braid or foil substrate. Depending on the specimen, the system can be used as a conventional torsion pendulum or in the supported TBA mode. The substrate generally used in a TBA experiment is a loose, heat-cleaned glass braid containing about 3600 filaments. The specimen is prepared by simply dipping the braid into a neat liquid or into a solution of the material of interest dissolved in a solvent.

Quantitative values of the shear and loss moduli of polymer films are obtained from the natural frequency (~ 1 Hz) and damping of the induced oscillations of the torsion pendulum. Some advantages of using the torsion pendulum for the dynamic mechanical thermal analysis of polymers are its fundamental simplicity, its relatively low cost, its utilization of the pendulum's resonance frequency (which means that the measurement is always made at the maximum sensitivity, since the specimen is a primary component of the pendulum), its use of a non-drag optical transducer which permits the use of small specimens (< 10 mg), and its free movement (because the pendulum is fixed only at one end, no adjustments for thermal expansion or contraction are required). In the torsional braid analysis (TBA) mode, a glass braid (or other substrate) is impregnated with a polymer solution or polymer "melt", and measurements are made on the composite specimen. Although the specimen's relative rigidity is obtained instead of its shear modulus, the transition temperatures can readily be identified. The advantages of the TBA technique include ease of specimen fabrication, vertical self-alignment (by gravity), and the capability of monitoring the physical properties of a specimen from the liquid to the solid state (as in the cure by chemical reaction of a thermosetting resin, or on cooling through the glass transition region of a thermoplastic system). For isothermal kinetic studies, the apparatus can be preset at a given temperature and the specimen can be lowered to the correct position within a few seconds.

The pendulum is intermittently set into oscillation to generate a series of damped waves as the material properties of the specimen change with temperature and/or time.

The shear modulus, G', is given by

$$G' = KI(2\pi f)^2 \left[1 + \left(\frac{\Delta}{2\pi}\right)\right] \qquad [6]$$

where f is the frequency (Hz) of the oscillation, I is the moment of inertia, Δ is the logarithmic decrement [$\Delta=\ln(A_i/A_{i+1})$, A_i is the amplitude of the ith oscillation], and K is a geometric constant. Equation 6 is usually approximated by

$$G' \cong KI(2\pi f)^2 \quad [7]$$

when $\Delta/2\pi < 0.1$. The loss modulus is given by

$$G'' = \frac{KI\Delta}{\pi}(2\pi f)^2 \quad [8]$$

The loss tangent tan δ = G"/G', where δ is also the phase angle between the cyclic stress and cyclic strain vectors; it is a characteristic measure of the ratio of the energy dissipated per cycle to the maximum potential energy stored during a cycle.

$$\frac{G''}{G'} = \tan\delta \cong \frac{\Delta}{\pi} \cong \frac{\alpha}{\pi f} \quad [9]$$

where α is the damping coefficient. During a dynamic mechanical thermal analysis experiment, the moduli can be monitored by observing the changes in the frequency and logarithmic decrement (Δ) of the oscillations. In a torsion pendulum experiment, where the geometric constants are known, absolute moduli are obtained: [e.g., for a rectangular film,

$$G' = \frac{(2\pi f)^2 IL}{N}\left[1 + \left(\frac{\Delta}{2\pi}\right)^2\right] - \frac{mgb^2}{12N} \quad [10]$$

in which the form factor, N, is

$$N = \frac{a^3 b}{3} \left(1 - 0.63 \frac{a}{b}\right) \qquad [11]$$

where a, b, and L are the thickness, width (a < b/3), and length respectively of the film, g is the graviational constant, and m is the mass. The moment of inertia of the oscillating parts of the pendulum, I, can be obtained in a calibration experiment using a wire of known modulus and dimensions]. Although the geometric constants are not known in a TBA experiment, the frequency can be used to calculate a relative modulus, since $G' \propto f^2$. The TBA plots display the relative rigidity as $1/P^2$ where P is the natural period of the oscillations.

The natural frequency of the oscillations range from 0.05 Hz (for a supported fluid) to 10 Hz (for a solid), and the logarithmic decrement ranges from 0.001 to 3.0.

The data acquisition system (25) consists of a desktop computer (with printer, plotter, and disc drives), multiplexed switch, and a digital voltmeter, as shown in Figure 6. The computer monitors the temperature and torsion pendulum oscillation signals, controls the initiation of the oscillations, and derives the frequency and logarithmic decrement of the oscillations from the raw data. The reduced data are stored and are available for plotting, printing and further manipulation.

The computer controls the pendulum in a repetitive sequence, as shown in Figure 7: after monitoring the wave until it has damped out to a preset level, the computer aligns the pendulum for linear response of the transducer by rotating one of the polarizers so that the light beam passes through the center of the linear region, and then initiates new oscillations by cocking the pendulum against a spring (via a gear train), again monitoring the wave until the oscillations have decayed, and then releasing it so that the oscillations are centered about the center of the linear region. The wave is digitized and stored. The frequency and logarithmic decrement are extracted from the digitized wave by, for example, a non-linear least squares fit to the solution of the differential equation of motion:

$$\theta = \theta_o e^{-\alpha t} \cos(2\pi f t + \phi) + Bt + C \qquad [12]$$

where θ is the angular deformation of the pendulum as a function of time t, θ_o is the initial deformation, ϕ is the phase angle, B is the drift coefficient, and C is the offset.

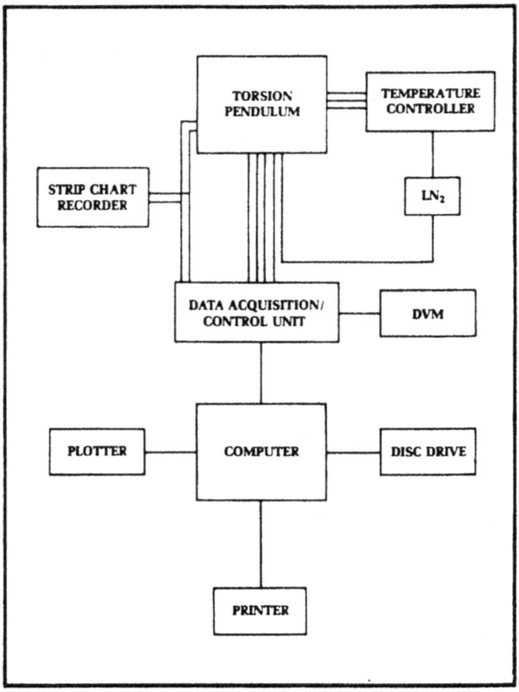

Figure 6. System Diagram

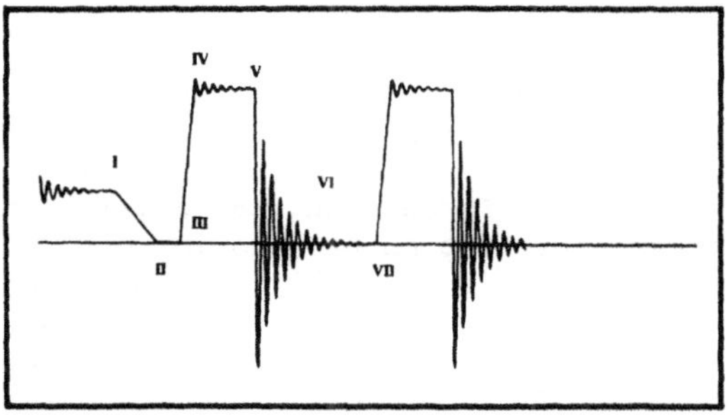

Figure 7. Theory of Operation. An analog electrical signal is obtained from a light beam passing through a pair of polarizers, one of which oscillates with the specimen. The pendulum is aligned for linear response and oscillations are initiated by the computer which also processes the damped waves to provide the dynamic mechanical spectrum. Automated torsion pendulum: control sequence. Key: I, previous wave decays, drift detected and correction begins; II, reference level of polarizer pair reached; III, wave initiating sequence begins. IV, decay of transients; V, free oscillations begin; VI, data collected; and VII, control sequence repeated.

APPLICATIONS

Much of the text of this review can be obtained by analysis of TBA data. This includes the TTT cure diagram itself (Fig. 1 and Fig. 4), and the T_g (from TBA) versus conversion (e.g., from calorimetry) relationship [Fig. 2]. Typical TBA plots for a) the isothermal cure of a liquid epoxy and b) the temperature dependent behavior of a cured epoxy are included in Figures 8 and 9, respectively. Transition times (e.g., vitrification and macroscopic gelation) and transition temperatures (e.g., T_g) are located by maxima in the mechanical damping (logarithmic decrement) curves. Quantitative TP data for sequential temperature scans of an initially undercured epoxy film are shown in Figure 10. (Note that postcure during the heating scan results in a higher T_g but a lower RT modulus in the post-cured material. This parallels the anomalous decrease in RT density which occurs with increased cure, as discussed earlier.) Reviews of the technique and its application to polymers in general have been published (1,25).

ACKNOWLEDGEMENT. Financial support has been provided by the Office of Naval Research

REFERENCES

1. J. K. Gillham in J. V. Dawkins, ed., Developments in Polymer Characterisation, Vol. 3, Applied Science Publishers, Ltd., London, 1982, Chap. 5, pp. 159-227.

2. J. K. Gillham, Polym. Eng. Sci. 26, 1429 (1986).

3. P. J. Flory, Principles of Polymer Chemistry, Cornell University Press, Ithaca, N.Y., 1953.

4. P. G. Babayevsky and J. K. Gillham, J. Appl. Polym. Sci. 17, 2067 (1973).

5. L. C. Chan, J. K. Gillham, A. J. Kinloch, and S. J. Shaw, Adv. Chem. Ser. 208, 235 (1984).

6. Ibid., 261 (1984).

7. S. Wu, Polymer 26, 1855 (1985).

8. J. K. Gillham, P. N. Reitz and M. J. Doyle, Polym. Eng. Sci., 8, 227 (1968).

9. M. B. Roller, Polym. Eng. Sci. 10, 692 (1979).

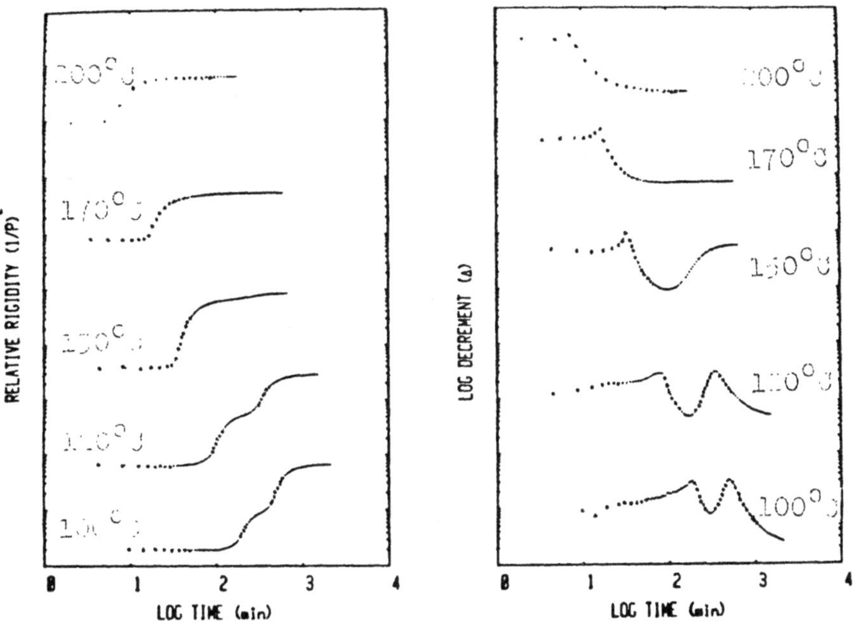

Figure 8. Representative TBA Isothermal (100 to 200°C) Cure Spectra of a Rubber-Modified Epoxy System (13). Relative Rigidity (Left) and Logarithmic Decrement (Right) versus Log_{10} Time.

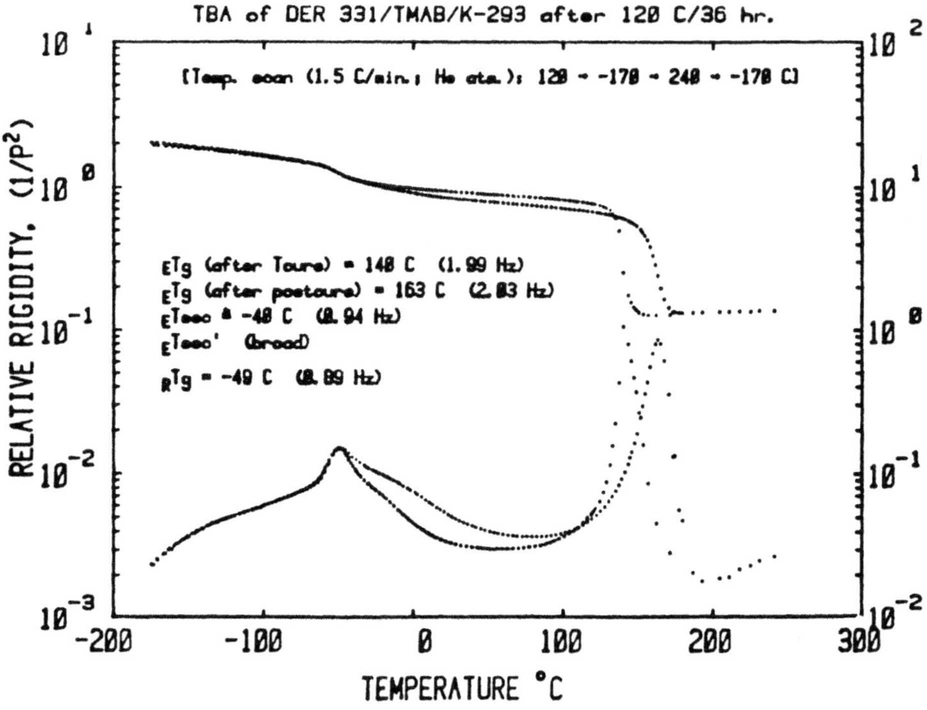

Figure 9. Sequential TBA Temperature Scans after Isothermal Cure (120°C/36 hr) of a Rubber-Modified Epoxy system (13). Temperature scan: 120 to -170 to 240 to -170°C at 1.5°C/min in dry helium atmosphere. Transition temperatures (°C) and frequencies (Hz) are marked.

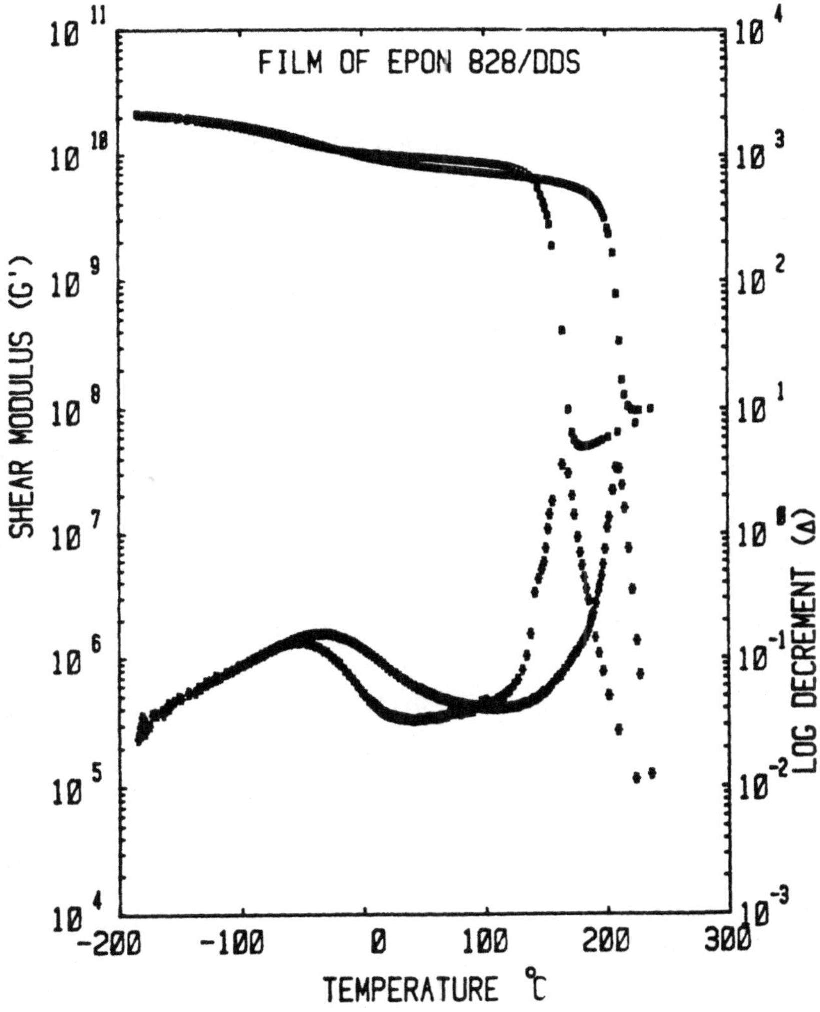

Figure 10. Sequential TP Temperature Scans of an Epoxy Film. Temperature scan 120 to -180 to 240 to -180°C at 1.5°C/min in dry helium atmosphere. Note that postcure during the heating scan results in a higher T_g but in a lower modulus at RT.

10. G. Palmese and J. K. Gillham, J. Appl. Polym. Sci. 34, 1925 (1987).

11. J. B. Enns and J. K. Gillham, Adv. Chem. Ser. 203, 27 (1983).

12. X. Peng and J. K. Gillham, J. Appl. Polym. Sci. 30, 4685 (1985).

13. L. C. Chan, H. N. Nae, and J. K. Gillham, Polym. Prepr. Am. Chem. Soc. Div. Org. Coat. Plast. Chem. 48, 566 (1983); J. Appl. Polym. Sci. 29, 3307 (1984).

14. J. D. Ferry, Viscoelastic Properties of Polymers, 3rd ed., John Wiley & Sons, Inc., New York, 1980.

15. J. B. Enns and J. K. Gillham, J. Appl. Polym. Sci. 28, 2567 (1983).

16. H. E. Adabbo and R. J. J. Williams, J. Appl. Polym. Sci. 27, 1327 (1982).

17. M. T. Aronhime and J. K. Gillham, J. Coat. Technol. 56(718), 35 (1984).

18. J. K. Gillham, J. A. Benci and A. Noshay, J. Appl. Polym. Sci. 18, 951 (1974).

19. M. T. Aronhime, X. Peng, J. K. Gillham and R. D. Small, J. Appl. Polym. Sci. 32, 3589 (1986).

20. K. P. Pang and J. K. Gillham, J. Appl. Polym. Sci. 37, 1969 (1989).

21. A. F. Lewis, M. J. Doyle and J. K. Gillham, Polym. Eng. Sci. 10, 683 (1979).

22. M. T. Aronhime and J. K. Gillham, J. Appl. Polym. Sci. 29, 2017 (1984).

23. S. Gan, J. K. Gillham and R. B. Prime, J. Appl. Polym. Sci. 37, 803 (1989).

24. G. Wisanrakkit, Ph.D. thesis, Princeton University, 1990.

25. J. B. Enns and J. K. Gillham, in T. Provder, ed., Computer Applications in Applied Polymer Science, Am. Chem. Soc., Symposium Series, No. 197, 329 (1982).

DYNAMIC SHEAR COMPLIANCE OF POLYMER MELTS AND NETWORKS

W. PECHHOLD, O. GRASSL AND W. v.SODEN
Abteilung Angewandte Physik
Universität Ulm, 7900 Ulm, FRG

ABSTRACT:

The mechanical relaxation processes in polymer melts and networks are discussed. This is done by decomposition of master curves of the dynamic shear compliance into (i) the glass relaxation with its plateau compliance J_{eN}, (ii) the shearband process, the relaxation strength ΔJ_B of which is reciprocal to the total crosslink density p_c and (iii) the flow relaxation ΔJ_F and viscous flow (for uncrosslinked melts only). The plateau compliance J_{eN} is exponentially reduced only by effective crosslinks ($p_c^* \approx p_c/30$). This behaviour is understood on the level of a meander superstructure, including shearbands. The observed saturation in J_{eN} at higher DCUP-crosslinking - which doesn't appear with radiation - can be explained by the lack of chemically induced effective crosslinks across the interfaces between meander cubes (because of larger shear fluctuations in these areas). Whereas crosslinking changes strongly the form and position of the master curve, a filler content reduces the compliance but keeps its lgω-dependence about constant. Quantitatively the filler effect on the zero shear compliance is described on the level of the meander model and its coarse grain structure: the paraelasticity of a file of meanders across a coarse grain of diameter D_{CG} is blocked up, if it contains at least one filler particle (of diameter D_F).

Therefore $J(\Phi_F)/J(0) = (1-\Phi_F)^{D_{CG}/D_F}$, the probability of finding no particle in a file (Φ_F total volume fraction of filler). Analyses on 4 series of filled cis-1,4-poly(isoprenes) can be accounted for by assuming the size of the (hypothetical) coarse grains of the polymer to be some 300nm. Also the prediction (by the simple formula), that smaller filler particles should be more effective in shear compliance reduction, is most probably valid.

O. Güven (ed.), Crosslinking and Scission in Polymers, 199–222.

1. Introduction

The most remarkable property of polymer melts or networks is rubber elasticity, which can only be explained by assuming paraelastic segmental movements yielding entropic restoring forces. In order to develop a detailed model theory, it is most helpful to analyse the (low stress) complex shear compliance $J(\omega,T)$ in the frame of suitable molecular models. For melts and amorphous networks, the frequency/temperature superposition is valid, as a rule, yielding master curves $J(\omega,T_o)$ and the proper shift factors on both axes. In the first part of this paper some results of master curve analysis on polystyrene fractions and binary mixtures in the melt are summarized and interpreted in the meander model [1]. Next, the strong influence of crosslinking density on the form and position of the master curve is discussed, on radiation- as well as DCUP-crosslinked polyisoprene series as examples. Finally it will be shown, that swelling or filler content do hardly change the form but strongly the position of the master curve.

2. The master curve and its analysis

Fig.1, as an example, shows the master curve for a commercial polyisobutylene (PIB B15, M_v=83000 in isooctane, $M_w/M_n \approx 2$). It has been composed of six J',J"-curves (measured in a double sandwich sample holder automatically in the frequency range 10^{-4} to 200 Hz at 150, 80, 0, -40, -55 and -75 °C)[1) by shifting them along both logarithmic scales. Curve fitting in the overlapping regions was simultaneously performed for J', J" by the computer. The vertical shift factors agree with the prediction $J',J'' \propto 1/\rho T$. The horizontal shift factors have been used to construct the activation diagramm (Fig.2, the 3 curves from the bottom).

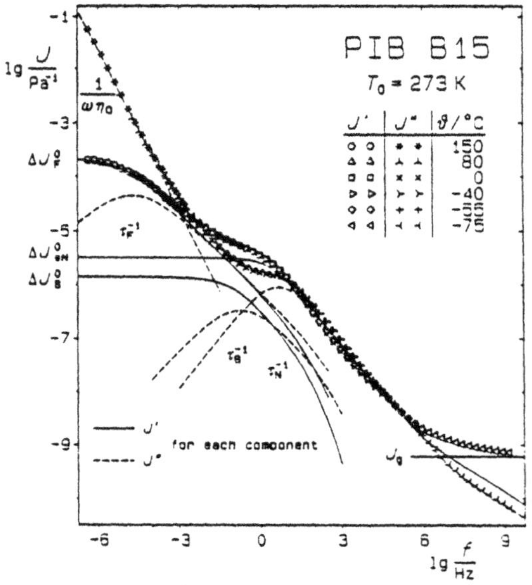

Fig. 1: Shear compliance - master curve for PIB B15 as deduced from measurements over 6 decades at 6 temperatures. Decomposition into three relaxation processes and viscous flow.

---

1) The measurements with the mechanical spectrometer (10^{-4}-200 Hz) including different sample holders and other details of the device are described in reference [2].

For a quantitative discussion of the master curves one may try their decomposition into a few components, which may be represented by Fuoss-Kirkwood[2] or Cole-Cole[3]-functions. A high quality fit[4] was achieved for all commercial PIBs investigated [3] by choosing a decomposition of the shear compliance into

J_g	the compliance of the glass,
ΔJ^o_{eN}, τ^{-1}_N, a_N	the glass relaxation process,
ΔJ^o_B, τ^{-1}_B, a_B	the shearband relaxation,
ΔJ^o_F, τ^{-1}_F, a_F	the flow relaxation and
$1/\omega\eta_o$	the loss-compliance due to viscous flow.

The ΔJ_i (relaxation strengths), τ_i (relaxation times) and a_i (widths of the distribution) as well as J_g and η_o were obtained by a least square fit of the mathematical functions [2,3] to the experimental master curve. Using these fit parameters, the 3 relaxation processes are indicated in Fig.1. Their relaxation frequencies f^i_{max} $(=1/2\pi\tau_i)$, obtained from this fit (at T_o=273K) and from the frequency shift factors, are plotted in the activation diagram (Fig.2).

For comparison, glass relaxation frequencies taken from G"- and ε"-maxima as well as ε''_R-maxima from dipolar label relaxation are included [1]. One single activation curve, fitted to the dielectric frequencies [1], accounts approximately also for the other processes after vertical shifting. This means that the rapid local (segmental) movement (G", ε"-maxima) is the elementary step of larger scale (multiple step) fluctuations, to be described next.

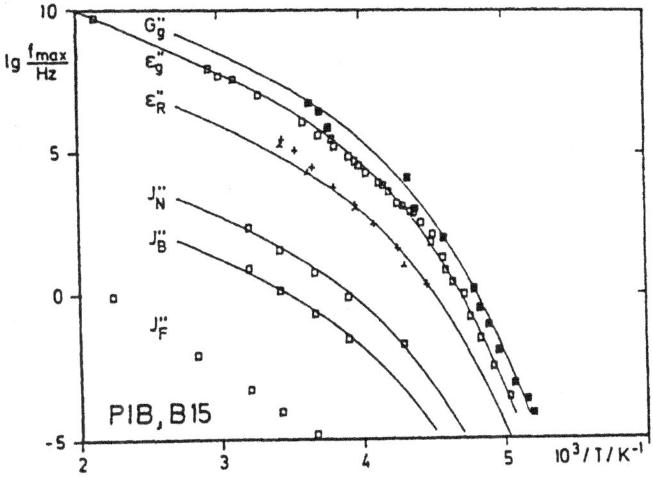

Fig. 2: Activation diagramm for PIB B15 from dielectric and mechanical measurements (cp. text).

---

2) $J''_i = a_i \Delta J_i / 2\cosh(a_i \ln\omega\tau_i)$; J' by Kramers-Kronig-relation

3) $J_i = \Delta J_i/[1+(j\omega\tau_i)^{a_i}]$ (complex analytical function)

4) The details of the decomposition procedure and the quality of the fit are also given in reference [2]

3. The meander superstructure and its shear deformation modes

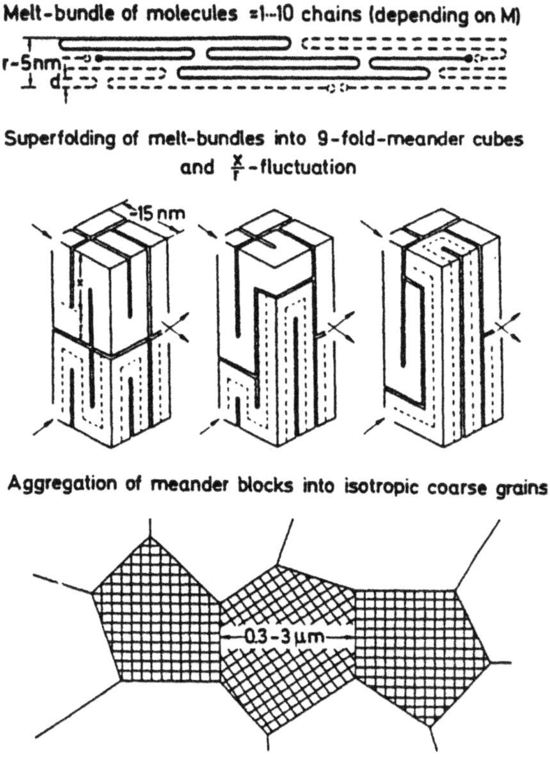

Fig.3: 3 levels of order in polymer melts or networks as described in the meander model

The meander model is quantitatively discussed in more detail in [1,4-8]. Its basic assumptions are (i) the cluster-entropy-hypothesis (CEH) and (ii) a short range order concept which favours chain parallelism across longer distances, i.e. the bundle concept. Adjusting to all energetically favoured conformations, the nearly parallel molecules are assumed to fold back and forth as defined by the free energy of a tight fold. The statistical tight folding and reptation of each molecule in the bundle results in a one-dimensional (quasi-parallel) coil appearance of its overall geometry (Fig.3, top). To gain orientational entropy of its segment lines (on the basis of CEH) a bundle (or topological tube) tightly superfolds, thereby changing its direction and becoming contracted.

The most simple and spacefilling topology of superfolding is an arrangement of meander cubes, linked together via space diagonals (serving as axes of meander cube rotation).

By minimization of the free energy per segment (including superfolding, shear fluctuation and rotation) the following results are obtained [1], which approximately read:

most probable 9-fold-meander cube, i.e. $x/r = 2$ (1)

equilibrium bundle diameter $r/d \approx (s/d)\Delta g_f^*/kT$ (2)

superfold free energy per segment $\Delta g_s^{sf}/kT \approx -d/3r$ (3)

Δg_f^*, the free energy of one superfold in a layer of molecules, gets reduced at lower M by substitution of a pair of chain ends for the tight chain fold within the superfold. Therefore r and the meander cube size decrease at lower M. Fluctuations in x/r and r/d have to be allowed for but must not violate topology. Relying on CEH, these fluctuations can take place by reversibly coupling of any two adjacent cubes in a suitable relative orientation (x/r-fluctuation, Fig.3,middle) and by changing the contour of the bundle cross-section without varying its area (r/d-fluctuation). The competition between the cube rotation entropy (the statistical element being a cube segment line) and the appropriate interaction of adjacent cubes may account for a liquid crystalline transition, in which the paracrystalline but isotropic grains (0.3 - 3µm, Fig.3, bottom) become nematically ordered [6,8].

In the meander model, the plateau compliance is explained by paraelastic intrameander shear deformation (Fig.5), mediated by dislocation movement (Fig.4), but being restricted (in the bulk) to 1d-displacements between adjacent layers of molecules in the cube ($\gamma_{max}^{intra} = 1/\sqrt{3}$).

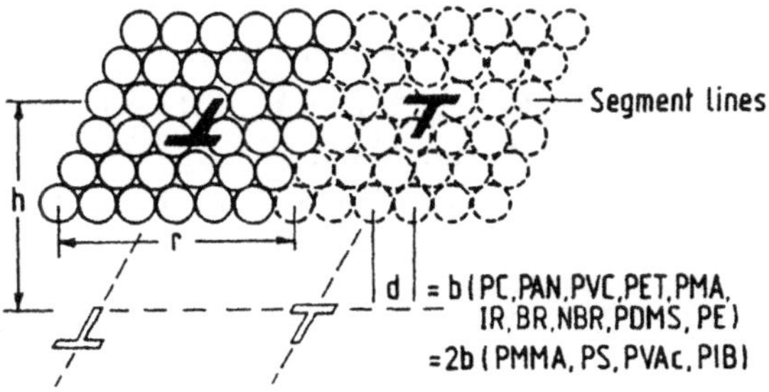

Fig.4: Dislocation wall concept of intrameander shear deformation. b, Burgers-vector; d, interchain distance.

To derive the cross-sectional shear compliance $2s_{1212}$ (3-axis in chain direction) under the assumption that only a fraction P of layers can be displaced by ±d/2 one has for the total $\gamma^{(12)}$

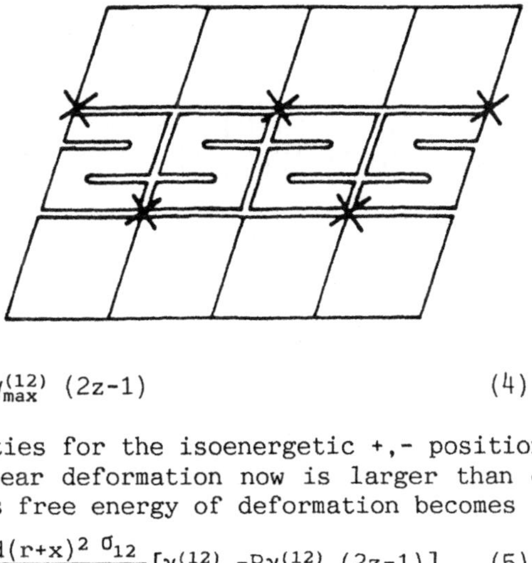

Fig.5: Intrameander shear deformation with $\gamma_{max}^{intra} = 1/\sqrt{3}$

$$\gamma^{(12)} = P\gamma_{max}^{(12)} [z-(1-z)] = P\gamma_{max}^{(12)} (2z-1) \quad (4)$$

with z, (1-z) as the probabilities for the isoenergetic +,- positions. The statistical element for shear deformation now is larger than one layer, $d(r+x)^2$, by 1/P, and its free energy of deformation becomes

$$\frac{\Delta g_{def}}{kT} = z\ln z+(1-z)\ln(1-z)+\frac{d(r+x)^2}{P}\frac{\sigma_{12}}{kT}[\gamma^{(12)} - P\gamma_{max}^{(12)} (2z-1)] \quad (5)$$

the last term takes into account condition (4), with a Lagrangian multiplier directly connected to the shear stress σ_{12} via its definition $\sigma = \partial(g/V)/\partial\gamma$.
Minimization of (5) for z yields

$$\gamma^{(12)} = P\gamma_{max}^{(12)} \tanh [d(r+x)^2 \gamma_{max}^{(12)} \sigma_{12}/kT] \quad (6)$$

In the limit of small amplitudes ($\sigma \to 0$) the anisotropic compliance becomes

$$2s_{1212} = 2s_{2121} = \partial\gamma^{(12)}/\partial\sigma_{12} = P(\gamma_{max}^{(12)})^2 d(r+x)^2/kT \qquad (7)$$

In a similar way the shear compliance parallel to the chain direction is derived [8], and reads

$$2s_{3131} = 2s_{3232} = \partial\gamma^{(31)}/\partial\sigma_{31} = \frac{2}{3}P(\gamma_{max}^{(31)})^2 d(r+x)^2/kT \qquad (7a)$$

With $\gamma_{max}^{(31)} = \gamma_{max}^{(12)} = 1/\sqrt{3}$ one gets for the isotropic case - using Reuss-averaging -

$$J_{eN}^{o} \approx \frac{1}{5}\left\{ \begin{matrix} s_{1212} + s_{2121} \\ + s_{3131} + s_{3232} \end{matrix} \right\} = \frac{d(r+x)^2}{9kT}P \qquad (8)$$

The plateau compliance (8) shows the 1/T-characteristic of entropy-elasticity and represents the most precise measure for the meander cube side length (r+x). The probability factor P, which, as a rule, is unity for melts, will become important for networks with medium or strong crosslinking.

In the glass relaxation process the plateau compliance gets reduced to the compliance of the glass. The relaxation frequency f_{max} (1/T) has been derived in [1,4,7], assuming an intramolecular activation energy Q_γ for the jumping segment (at the dislocation) as well as a cooperative (probability) factor for the existence of a dislocation wall structure. This expression for the activation curve reads

$$f_{max} = \frac{f_o}{\pi}\exp\left[-\frac{Q_\gamma}{kT}\right]\left[1-\left(1-e^{-\frac{\varepsilon_s}{kT}}\right)^{\frac{3r}{d}}\right]^{3\left(\frac{3r}{d}\right)^2\frac{d}{s}} \qquad (9)$$

and has been applied to the glass relaxation of many polymer [1,4,7], cp. Fig.2 for PIB.

So far, the intrameander shear, i.e. shear fluctuations - restricted to one chain distance per layer of molecules - of the meander cubes have been discussed. In equilibrium, they explain the plateau compliance (8), but they account only for a maximum shear $\gamma_{max}^{intra} = 1/\sqrt{3}$ (Fig.5). Beyond these fluctuations, large scale deformations become possible by interbundle displacement, i.e. by unfolding of suitably arranged meander cubes (Fig.6). To this purpose, whole files of meander cubes or even layers of them must

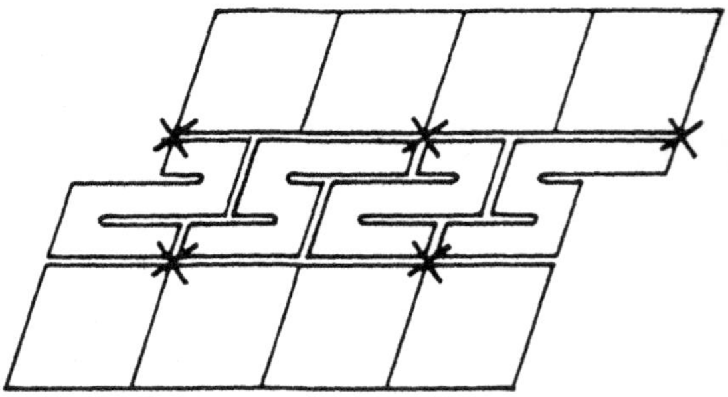

Fig.6: Interbundle shear as realized in a shearband (maximum deformation $\gamma_m = 9$)

cooperate in a shearband process, the stress-strain behaviour of which was first derived in [8].
This model theory was generalized in reference [1] taking into account rearrangements within the shearband which reduce its orientation correlation and therefore increase its compliance as well as its probability $\beta(\sigma)$: (i) only the polymer fraction Φ, which cannot recover its cube rotation mode by reptation (at the time scale of interest), may maintain its shearband orientation. (ii) Part of this fraction, $(1-\xi)$, may albeit rearrange (by movement of folds) and reduce the orientation (x_2-x_1) (Fig. 7b), the origin of the restoring force. For statistical networks a simple picture to estimate ξ will be used later on. Fig.7 may elucidate the meaning of Φ and ξ with 3 examples.

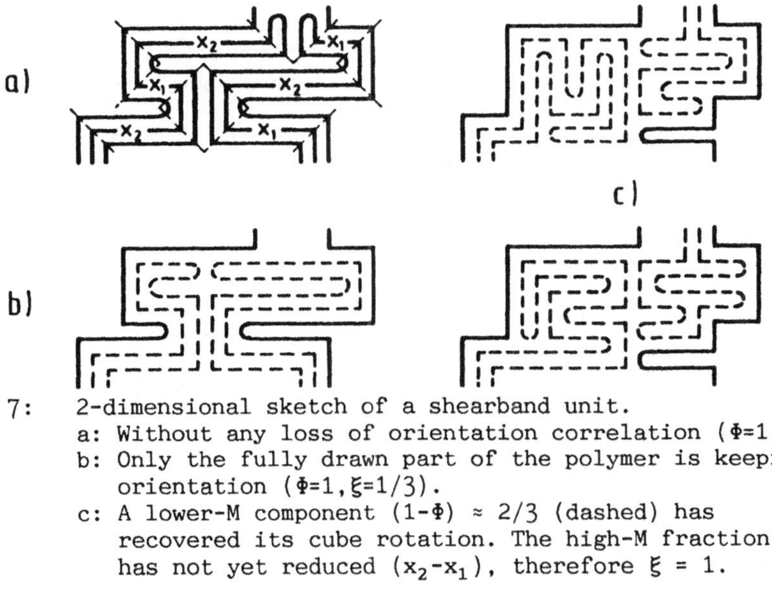

Fig. 7: 2-dimensional sketch of a shearband unit.
a: Without any loss of orientation correlation (Φ=1,ξ=1)
b: Only the fully drawn part of the polymer is keeping the orientation (Φ=1,ξ=1/3).
c: A lower-M component $(1-\Phi) \approx 2/3$ (dashed) has recovered its cube rotation. The high-M fraction (Φ) has not yet reduced (x_2-x_1), therefore $\xi = 1$.

The strength of the relaxation process due to shearband deformation can be written in the linear case after Reuss-averaging [1]

$$\Delta J_B^o = \frac{\beta(\Phi)}{\Phi\cdot\xi}\,\frac{s\cdot d(r+x)^2\gamma_m^2}{5xkT} \qquad (10)$$

$\beta(\Phi)$ is the shearband concentration, derived in [1, Equation 25], and amounts to some 2 per cent for dry rubber at $\sigma \to 0$.

Besides plateau compliance and shearband relaxation (i.e. rubber-elasticity), polymer melts show viscous flow and a simultaneous recoverable strain if the stress is released. In dynamic mechanical investigations, the latter corresponds to a paraelastic relaxation process taking place in the same frequency range as the onset of a $J''=1/\omega\eta$ behaviour, i.e. when chain reptation becomes effective.

In the meander model, this paraelastic deformation is attributed to an extra shear γ_r (Fig.8) caused by the reptating chains, when leaving one junction and entering a neighbouring one.

Fig. 8: 2-dimensional sketch of the paraelastic deformation on the upper interface of a shearband, caused by spread junctions.

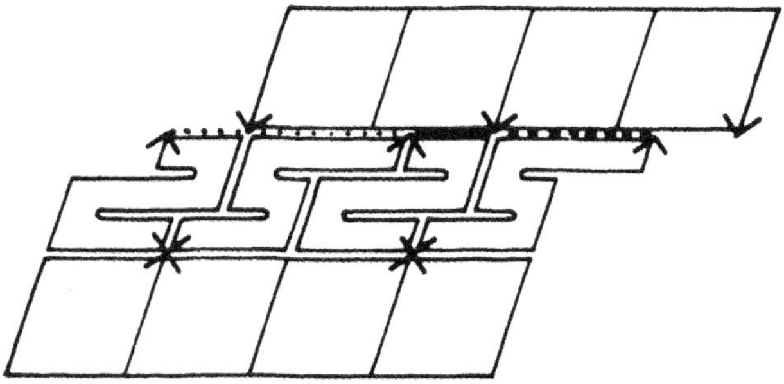

To minimize interfacial energy of these spread junctions, two assumptions are made:
(i) Only the upper and lower interface of a shearband is appropriate to adapt extra chain sections with a minimum of chain crossing.
(ii) The extra sections (fully drawn, dashed and dotted lines in Fig.8) are the shortest bridges between the respective junctions (with no back and forth folding). On this basis the flow relaxation strength has been derived [1]

$$\Delta J_F^o = \frac{2}{5}\beta \cdot 216 \frac{qd^2 r/kT}{1+(q-1)p} F(Q,m) \tag{11}$$

with β shearband concentration, q average number of parallel stems/molecule, r bundle diameter, d interchain distance, l_o monomer length, p_f^o tight fold probability (per monomer), p probability of a tight fold to bridge a junction (Fig.9), M_o molecular weight of the monomer, $m=(l_o/6r)M/M_o$, $Q=\exp[-6rp_f^o/l_o]$. The analytical function F is given in [1].
To predict the zero shear rate viscosity

$$\eta_o = 2\tau_F/\Delta J_F^o \tag{12}$$

one has to derive the flow relaxation time τ_F. This can be done as follows: The flow relaxation requires disentanglement of shearband junctions and hence reptation of the chains across half of their lengths L on the average.
For fractions one may use the simple formula $x^2=2D\tau$, and gets for τ_F

$$\tau_F = (L/2)^2/2D_1 \tag{13}$$

D_1 being the curvilinear diffusion constant.

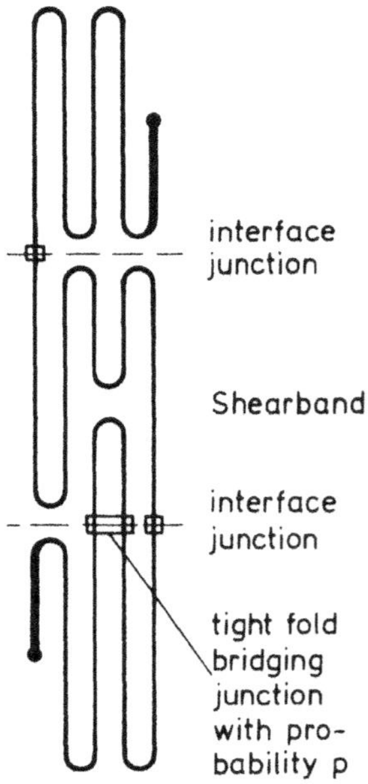

Fig.9: Shorthand sketch of a tight folded molecule, crossing two shearband interfaces.

To derive D_1 for chain reptation in the bulk in the frame of the meander model, one has to take into account that the chain folds are either sessile at the junctions (with probability 1-p) or fluctuating lengthwise (p) within the molecules (Fig.9). The latter motion doesn't contribute to an overall transport of the molecule's centre of mass (C.M.), if no diluent is present. Therefore the only segments to jump *independently* with frequency Γ_o are those at the ends of the molecule (outside the folds, Fig.9, fat drawn). Its average number per molecule is $l_o/p_f^o\lambda$, if λ denotes the length of the diffusing segment (in the meander model, the average length of the straight chain parts i.e. $\lambda \approx 2r$). The C.M.-displacement resulting from an independent segmental jump of length l_o (monomer length) is $a=\pm l_o\lambda/L$, and the curvilinear diffusion constant D_1 becomes (with $L=nl_o$)

$$D_1 = \frac{1}{2}a^2\Gamma = \frac{1}{2}\left(\frac{l_o\lambda}{L}\right)^2\frac{l_o}{p_f^o\lambda}\Gamma_o = \frac{\lambda l_o^2}{2L}\,\frac{\Gamma_o}{np_f^o} \qquad (14)$$

(14) indicates two regimes of M-dependence of D_1: (i) For $np_f^o \leq 1$ (molecules preferentially unfolded) np_f^o in (14) must be substituted by 1 and $D_1 \propto M^{-1}$. (ii) For $np_f^o > 1$, D_1 will be proportional to M^{-2}. So, $M_c = M_o/p_f^o$ signifies a critical molecular weight, which (using the p_f^o-data from [1]) comes close to the M_c-data from viscosity measurements [3]. Substitution of (14) into (13) reads

$$\tau_F = \left(\frac{L}{2}\right)^2\frac{L}{\lambda l_o^2}\frac{np_f^o}{\Gamma_o} = \frac{l_o np_f^o}{4\lambda\Gamma_o}n^3 = \frac{l_o}{4\lambda\Gamma_o}\left(\frac{M_c}{M_o}\right)^3\cdot\begin{cases}(M/M_c)^3 & \text{for } M\leq M_c\\ (M/M_c)^4 & \text{for } M>M_c\end{cases} \qquad (13a)$$

In the last expression, $np_f^o=1$ and $p_f^o=M_o/M_c$ were used for $M\leq M_c$ and $M>M_c$, respectively. Finally η_o can be calculated from (12), if ΔJ_f^o and its M-dependence is known (see below).

4. Viscoelasticity of polymer fractions and binary blends in the melt.

In Fig. 10a, shear compliance mastercurves for 10 aPS fractions of $M_w/M_n<1.1$ and with molecular weights beween 10^4 up to $2.4\cdot10^6$ g/mol are shown [1]. The reference temperature is T_o=405.5K in all cases and each pair of J',J"(f)-curves is shifted by one decade in the ordinate against the preceding one, starting at the bottom.
If one tries to analyze these curves by decomposing each pair into a few symmetric relaxation processes - characterized, e.g. by Cole-Cole distributions[3] - it turns out that a good fit can be achieved with two processes (glass relaxation and flow relaxation) and the viscous flow. This means that the shear band process between the glass and flow process is absent or lower in strength by at least a factor of 1/5 of the plateau compliance. Because all the chains in a fraction behave the same, it follows that Φ=1. One may also say ξ=1, because a rearrangement of strands is opposed to the chain reptation taking place within the bundles. Therefore, we have to apply Equation (10)

with Φ=1, ξ=1, to estimate the shearband relaxation strength for fractions. Assuming for aPS: $-\Delta g_{rot}/kT=2/3$, $\gamma_m=9$, $d=2s=4l_o=0.88$nm, $x/r=2$, $r_\infty/d=6$, we get $\beta(1)\approx0.018$ from [1, Eq.25] and from Equation (10) and Equation (8)

$$\Delta J_B^\infty = \beta(1)\frac{s\cdot d(r+x)^2\gamma_m^2}{5xkT} = \beta(1)\frac{9s}{5x}\gamma_m^2 J_{eN}^o \approx 0.11 J_{eN}^o \qquad (15)$$

which is indeed much smaller than the plateau compliance and cannot be detected by decomposition of the master curves.

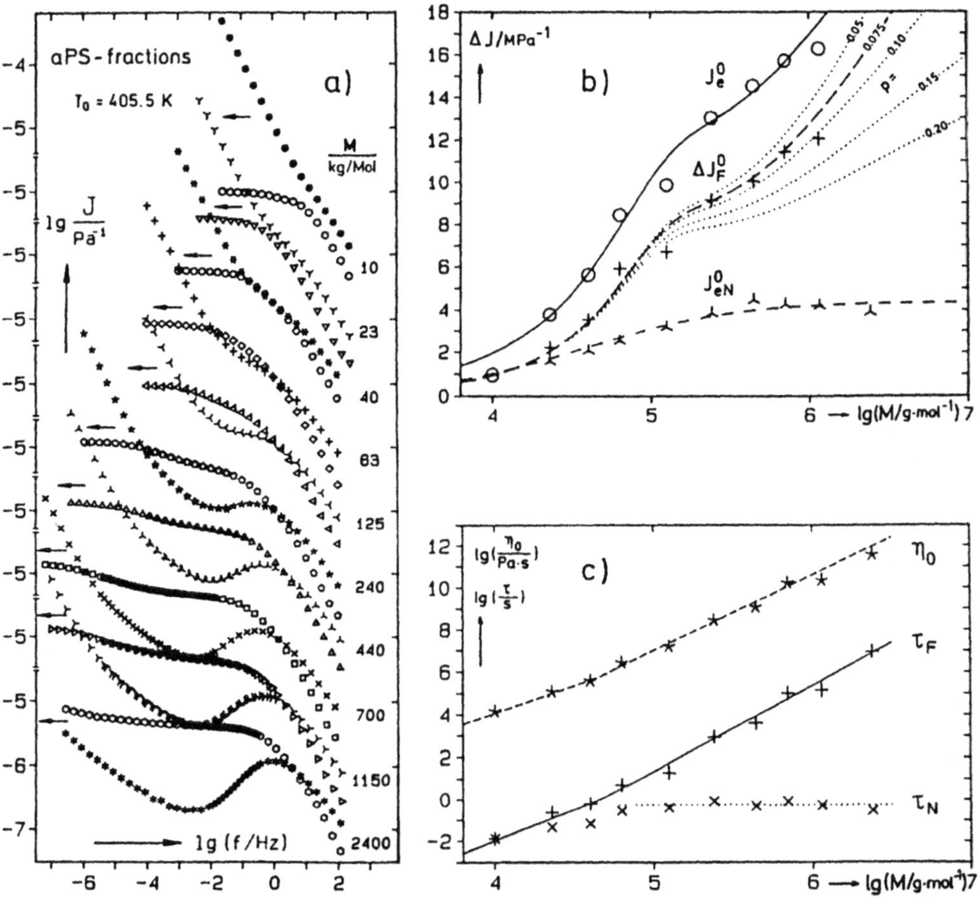

Fig. 10: (a) Compliance master curves of 10 aPS fractions in the melt. (b) Relaxation strengths of the glass and flow processes (determined by curve decomposition) in dependence on M. (c) Relaxation frequencies τ_N and τ_F and the zero-shear viscosity η_o (by the same analysis) in dependence on M.

For discussion of the glass and flow process, the respective relaxation strengths ΔJ^o_{eN} ($=J^o_{eN}$) and ΔJ^o_F are plotted (together with their sum, the steady state compliance J^o_e) versus lgM in Fig.10b and compared with the predictions of Equations (8) and (11), given by the dashed curves.
The agreement between theory and experiment is good within experimental error. The small value of 0.07 to 0.10 for the probability p in Equation (11) points out that most of the tight folds do not cross the junctions but prefer to stay near them. Because it is still in dispute whether the steady state compliance does level off at high M or continue to increase, one could think that it depends on the polymer and its probability p.
The decomposition of the master curves also yields the zero shear viscosity η_o and the relaxation times ($\tau_{N,F}=1/2\pi f_{N,F}$) and their M-dependence, plotted in Fig. 10c.
Whereas τ_N (the glass relaxation time for the compliance) stays constant for higher M (because $r \approx r_\infty$), τ_F shows the two regimes of M-dependence predicted in (13a). Approximately one can extract from Fig.10c

$$\tau_F/s \approx 0.65 \cdot \begin{cases} (M/M_c)^3 \\ (M/M_c)^4 \end{cases} \text{for } \begin{matrix} M < \\ M > \end{matrix} M_c \approx 4 \cdot 10^4 \text{g/mol} \qquad (16)$$

and, adapting a power law, from Fig.10b

$$\Delta J_F/\text{MPa}^{-1} \approx 3.5\ (M/M_c)^{0.5} \text{ for the whole M-range.} \qquad (17)$$

Inserting (16,17) into (12)[5)] the zero shear rate viscosity becomes

$$\eta_o/\text{MPa}\cdot\text{s} \approx 0.37 \cdot \begin{cases} (M/M_c)^{2.5} \\ (M/M_c)^{3.5} \end{cases} \text{for } \begin{matrix} M < \\ M > \end{matrix} M_c \qquad (18)$$

which nicely describes the experimental data (Fig.10c) and can also be considered as an experimental proof of the relation (12)[5)].
For some polymers (e.g. polyisoprene, polysiloxane) the viscosity of molten fractions below M_c follows $\eta_o \sim M$. This is the consequence, via (12), of $\Delta J_F \sim M^2$, observed in this case.

To lead to polymer networks, the shear compliance master curves of which shall be discussed next, it is worthwhile asking the question of whether the shearband process can be observed in polymer melts at all. The answer is definitely yes, if one inspects the master curves in Fig.11 as the fingerprints of binary aPS blends of different mixing ratios.

----

5)

To derive Equation (12), one may consider a flow element (serving one spreading junction) in Fig.8, and define a jump frequency Γ_F for flow. Then one has $\tau_F=1/2\Gamma_F$ ($\sigma \to 0$) and the two definitions $\sigma=\eta\dot{\gamma}$ and $\gamma=\Delta J_F \cdot \sigma$, with σ being that part of the shear stress inducing viscous flow. Accepting the plausible assumption $\dot{\gamma}=\gamma\Gamma_F$ for the shear rate, one gets by substitution the relation (12).

Since M_2/M_1 is about 11.3, the shorter chains can rearrange (dynamically fractionate by reptation) within the shearband, gaining back their free energy of rotation (cp. Fig.7c). Therefore, Equation (10) must be applied with $\Phi<1$. With a more sophisticated application of the formulae above and a proper averaging over all chain distributions, the strength and relaxation frequencies of all three processes and their dependences on mixing ratio can be understood. Postponing the theory to a subsequent paper it may only be mentioned here that a good fit to the M_w-dependence of the shearband relaxation strength is not achieved unless about 40 files of meander cubes operate in parallel.

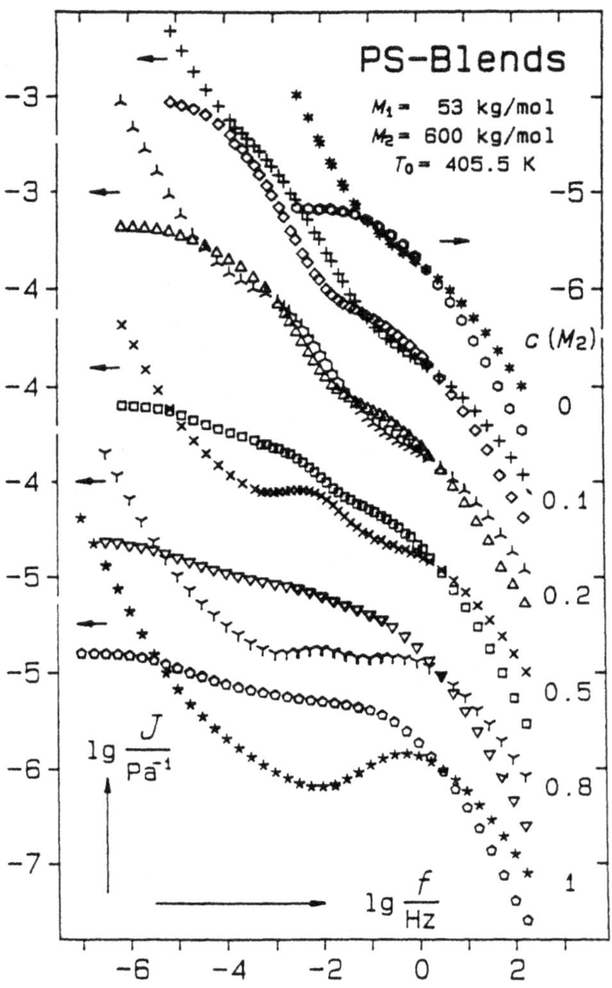

Fig. 11: Compliance master curves of 2 aPS fractions and 4 binary blends of them, the latter showing a rearranged shearband process in between the glass and flow relaxations.

5. The shear compliance of polymer networks in dependence on crosslink density.

The shear compliance of polymer networks is characterized by two relaxation processes: the glass process and a lower frequency relaxation which finally leads to the equilibrium compliance J_e. At low crosslinking, the low frequency process dominates in strength and is well separated from the glass process. With higher crosslinking it strongly decreases and accelerates until it seems to merge below the glass relaxation. This behaviour is common to end-linked model networks [9] as well as to statistical networks [1,10,11] and will be discussed in the last part of this section.

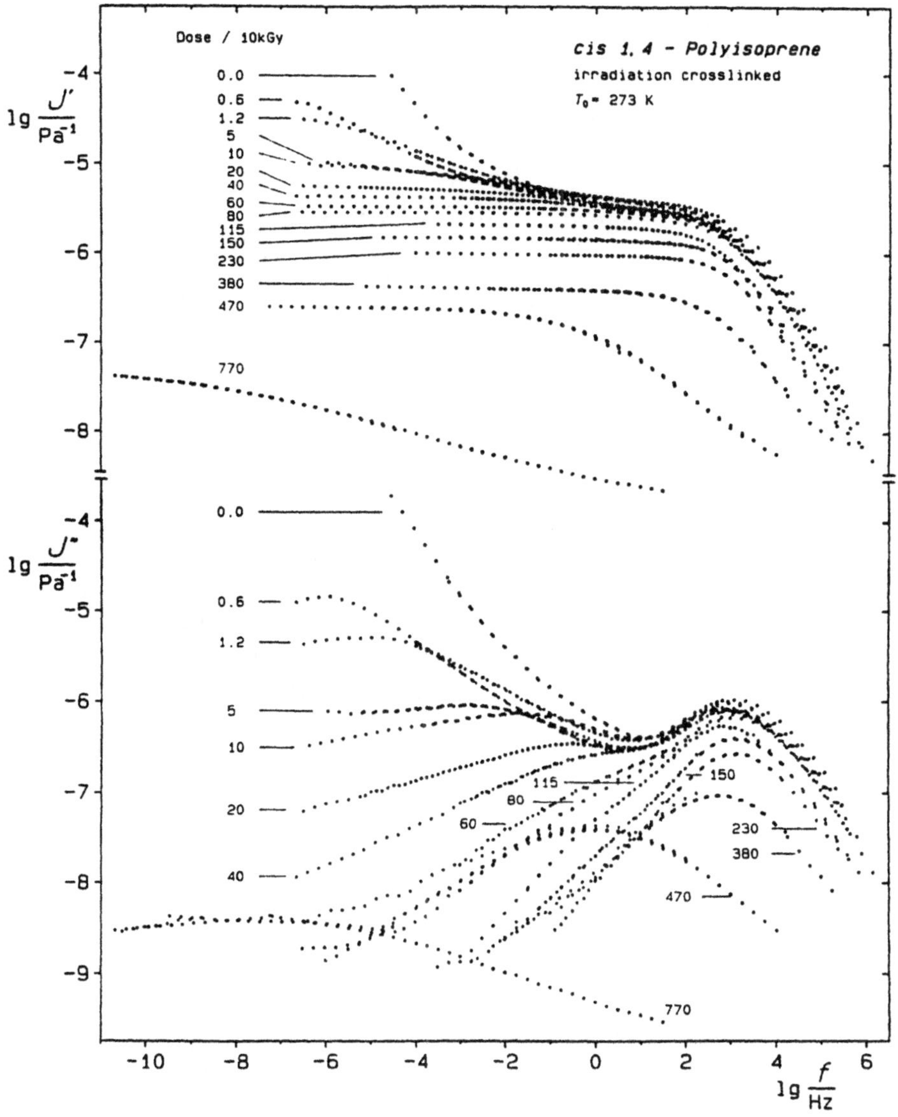

Fig. 12a: Compliance master curves of irradiation crosslinked IR 305 samples. The dose, in units of 10kGy, is assigned to each curve.

Figures 12a or 12b show master curves of cis-1,4-poly(isoprene)[6], crosslinked to various amounts by 2 MeV-electrons or by Dicumylperoxide (DCUP), respectively. Dropping a small flow component (for low crosslinking), most master curves could be perfectly decomposed into two relaxation processes, using Cole-Cole[3] distributions. The relaxation frequencies and relaxation strengths of these are plotted in Figures 13a,b and 14a,b, respectively.

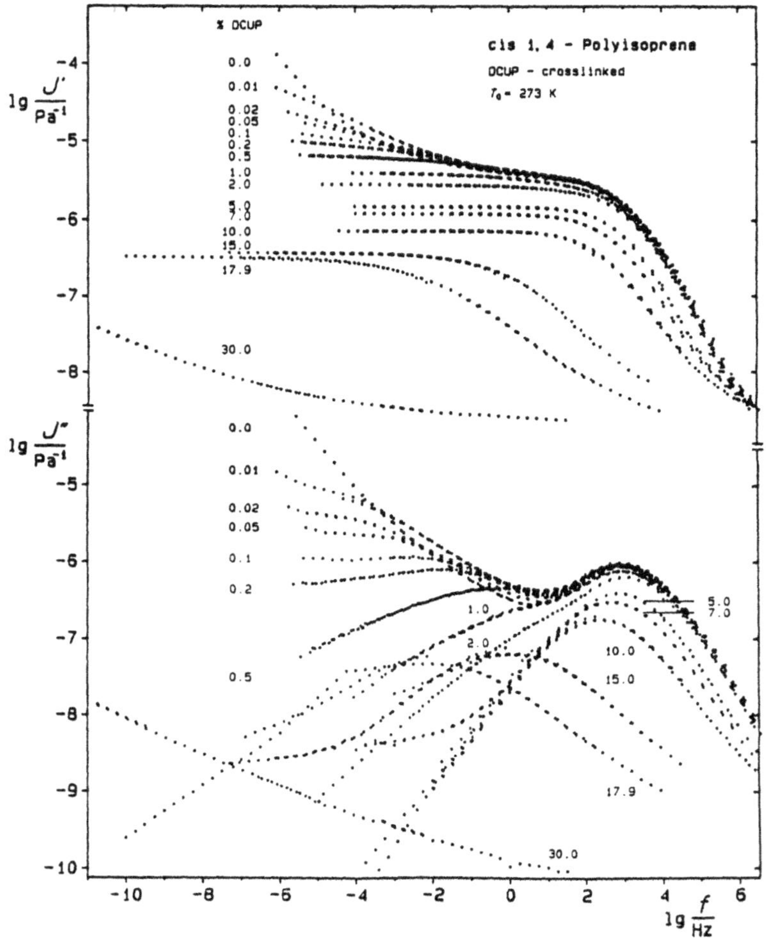

Fig. 12b: Compliance master curves of Dicumylperoxid (DCUP) cross-linked IR 305 samples. The phr of DCUP added to a sample before tempering it for 30min at 160°C is assigned to each curve.

---

[6] Cariflex IR 305(Shell), $M_w = 2 \cdot 10^6$, $M_n = 2.5 \cdot 10^5$ g/mol, anionically polymerized with Butyl-Li-catalyst, 93% cis-1,4, 7% 1,2. All samples were kindly provided by Dr. U. Eisele, Bayer AG, Leverkusen.

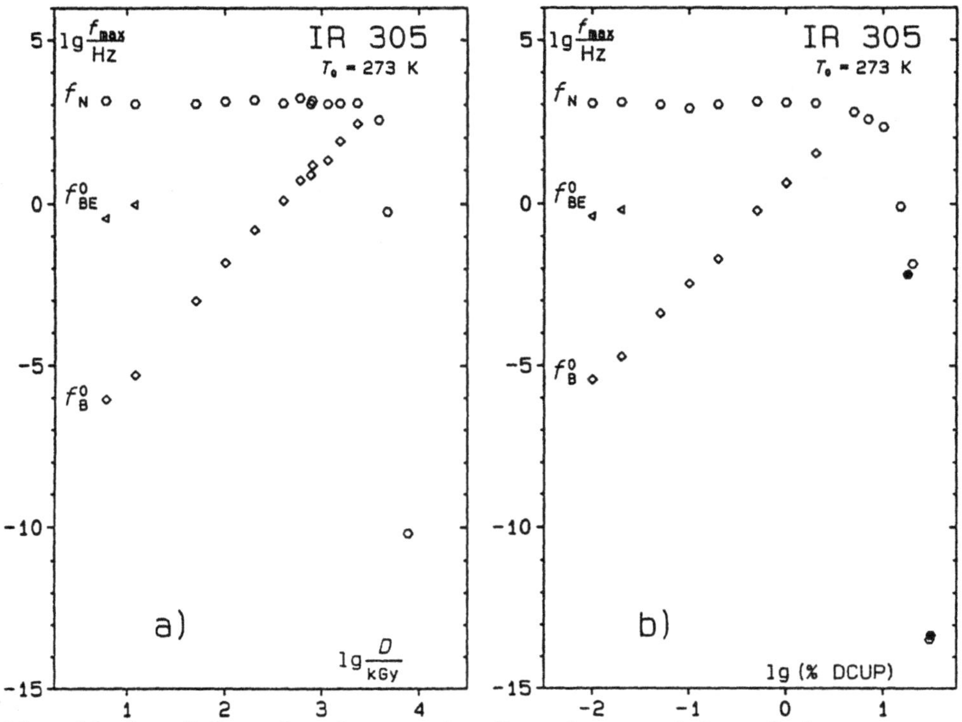

Fig. 13a,b: Relaxation frequencies from decomposition of the master curves in Fig.12 (f_N glass relaxation, f^o_B rearranged shearband process)

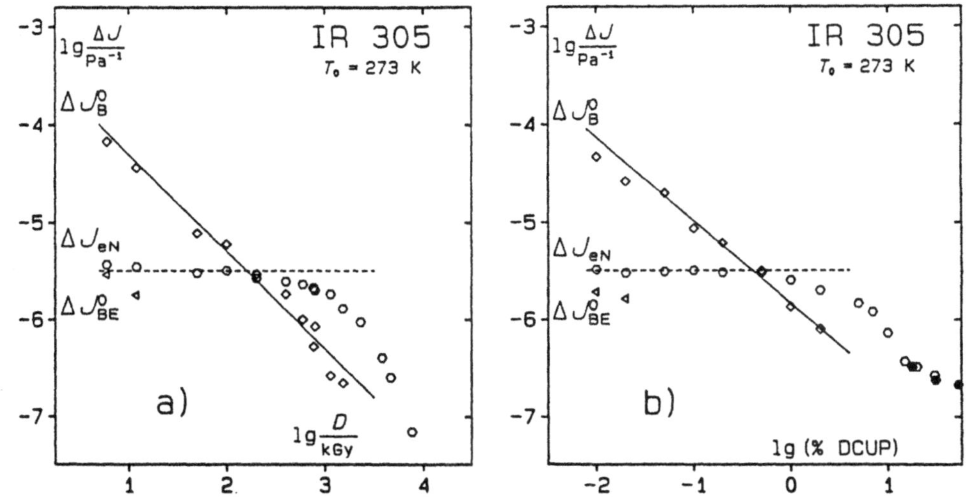

Fig. 14a,b: Relaxation strengths from decomposition of the master curves in Fig.12 (ΔJ_{eN} glass relaxation, ΔJ^o_B rearranged shearband process).

In the following we discuss these results (Fig.13a,b and 14a,b) separately for the glass relaxation and the shearband process.

The glass relaxation process

The glass relaxation frequency f_N (Fig.13) becomes influenced by crosslinking only at very high dose or DCUP addition (in the latter case, the cumenol formed during the reaction had been evaporated at 110°C, to avoid its swelling effect). This final influence probably can be understood by only statistical formation of adjacent crosslinks which strongly delay dislocation movement. The plateau compliance ΔJ_{eN} (Fig.14) starts to decrease very smoothly with crosslinking; its decay becomes very impressive if plotted semilogarithmically, as in Fig.15a,b. For its explanation one again must turn to the intrameander shear mode and its basic formula (8).

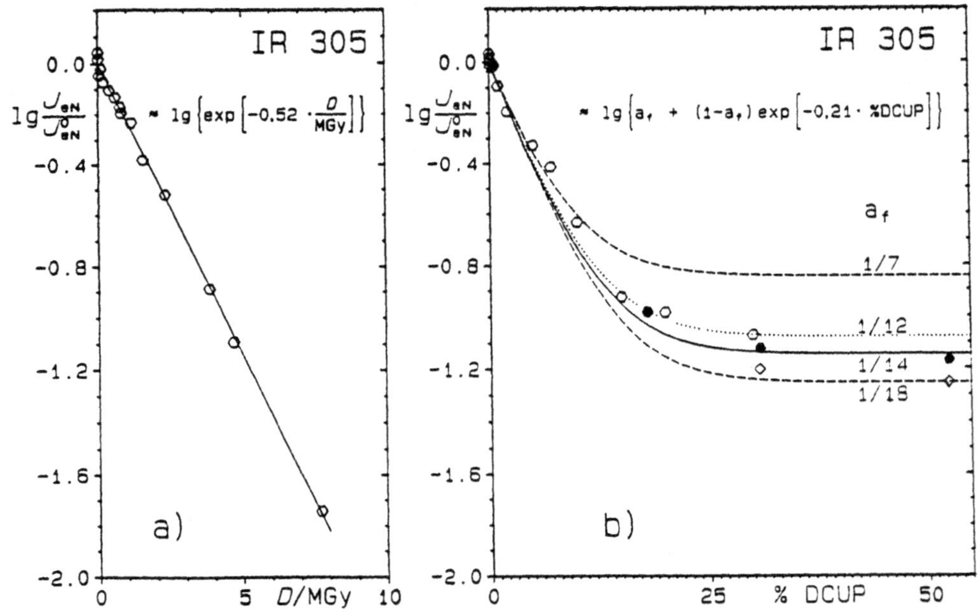

Fig. 15a,b: The reduced plateau compliance of IR-networks (plotted semilogarithmically) in dependence on the amount of crosslinking agent. (a) for radiation dose in MGy, (b) for DCUP in phr. For the meaning of the symbols cp. text.

If chemical crosslinks are present, say p_c per polymer segment, they will bridge adjacent layers of molecules in the meander cube and possibly hinder their relative displacements. But only a fraction p_c^* of the crosslinks present will effectively block the one d-displacement in the intrameander shear, as is illustrated in Fig.16 (for the cross-sectional shear): in the first two lines, uneffective crosslinks (open symbols) are shown, which can always be formed and which do not block (21)- and (12)-shear deformation. The third type of crosslink (filled symbol) can only be formed in one position and will block subsequent shear deformation (otherwise it must be broken). From this scheme, one concludes a geometrical probability $p_c^*/p_c = 1/7$ for a given crosslink to be effective. But the realistic p_c^*/p_c will be smaller, because of different crosslinking reaction probabilities. These should be larger for the ineffective crosslinks (between chains remaining in contact during shear deformation) than for the effective crosslinks (between chains being in contact only for short times).

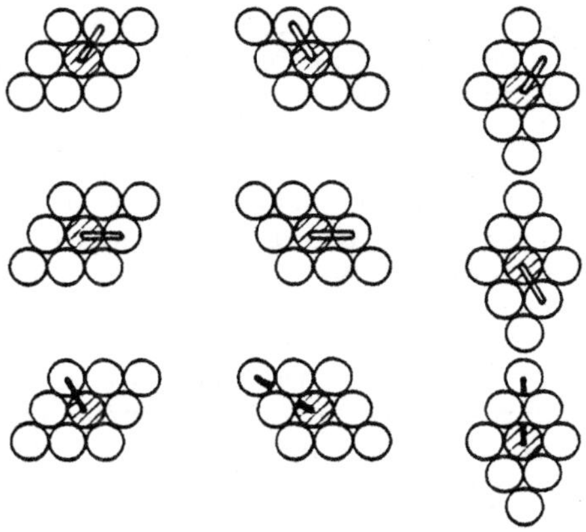

Fig.16: Ineffective (open) and effective (filled-in) crosslinks for cross-sectional shear.

Taking the ideal contact probability (without shear fluctuations) as a measure of efficiency, one gets as an upper limit

$$p_c^*/p_c \leq 1/7 \qquad (19)$$

To consider how effective chemical crosslinks influence plateau compliance, Equation (8) has to be applied, with P specified. It is reasonable that P, the probability of displaceable layers is equal to $P(p_c^*)$, the probability of finding no single effective crosslink between adjacent layers of molecules (consisting of $m=(r+x)^2/d\cdot s$ segments). Using Poisson distribution, one may write

$$P(p_c^*) = \exp(-mp_c^*). \qquad (20)$$

Evaluating Figures 15a,b, and assuming $3r/d\approx21$, $d=0.6$nm, $s=0.405$nm, i.e. $m=(3r)^2/d\cdot s\approx650$, one concludes

a) for irradiated IR: $p_c^*=8.0\cdot10^{-4}$ D/MGy throughout,
b) for DCUP-crosslinked IR: $p_c^*\approx3.2\cdot10^{-4}$ DCUP/phr,

in the range up to 3 per cent of DCUP. The ratio of effective crosslink densities for these two crosslink agents is

$$\frac{p_c^*(\mathrm{DCUP/phr})}{p_c^*(\mathrm{Dose/MGy})} = 0.4\frac{\mathrm{MGy}}{\mathrm{phr}} \qquad (21)$$

Whereas for irradiation crosslinking, Equation (20) with $p_c^* \propto D$ is valid at all doses examined, a saturation in J_{eN} is observed at higher DCUP-crosslinking. Only to a small amount, this may be due to inefficient crosslinking at high DCUP-content: The filled in hexagonal symbols, got from a two step crosslinking (after swelling a 5 p.c. sample with DCUP/hexane and drying at room temperature), are only a little lower (Fig.15b) as those for the samples (open hexagons) crosslinked as usual. A possible explanation for the obvious saturation effect in the frame of the meander model may be the lack of chemically induced effective crosslinks across the interfaces between meander cubes. This lack (or inefficiency) will be caused by larger shear fluctuations in these areas which increase the interchain distance across the interface and/or reduce the time of contact for the chemical agent. Quantitatively, the square-symbols (Fig.15b) indicate the reduction of the plateau compliance at 80°C, which levels off at a ratio $J_{eN}/J_{eN}^{o} \approx 1/18$, that would correspond to a meander cube size $3r/d=18$, if all interfaces between the cubes are free of effective crosslinks. The hexagonal symbols refer to compliances from decomposition of the master curves (i.e. at 0°C). The top curve in Fig.15b, which best fits the ratios measured on medium crosslinked samples, levels off at $J_{eN}/J_{eN}^{o} \approx 1/7$. This would correspond to a situation, where additionally all interfaces between the tight folded bundles within the cube are free of effective crosslinks (because of favoured shear fluctuation). It seems, that these interfaces of mostly parallel chains, become reduced in shear fluctuation with further crosslinking.

The saturation in J_{eN} with higher DCUP-addition does not indicate, however, a stagnation of further crosslinking. On the contrary, the glass relaxation starts to shift strongly to lower frequencies (Fig. 12b), very similar to that of the irradiated samples (Fig.12a).

But how could one explain this nonlinear frequency shift? Whereas single *ineffective* crosslinks (open symbols in Fig.16) are assumed to switch freely in intrameander shear, neither affecting dislocation movement nor increasing the free energy of formation ε_s of the segment dislocation, this obviously will change, if a segment is tied to two adjacent layers of molecules by two ineffective crosslinks. In this case of a double crosslink the local ε_s will be strongly increased (by additional intra- and intermolecular deformation energy). Since a dislocation wall contains many segment dislocations, the average ε_s will continously increase proportional to p_c^2, if the double crosslinks are formed statistically. In Figure 17a,b the activation curves for both crosslinking series are drawn according to formula (9), varying only the parameter ε_s. The fit to the experimental data looks very good and moreover $(\varepsilon_s - \varepsilon_{s0})$ turns out to change proportional to p_c^2, at least for $p_c < 0.2$. A similar dependance should be valid for the glass temperature T_g. Further experiments on sulfur-crosslinked IR are expected to show $T_g \sim p_c$ [15] and will shine new light on the strong T_g-shift in the case of sulfur vulcanization.

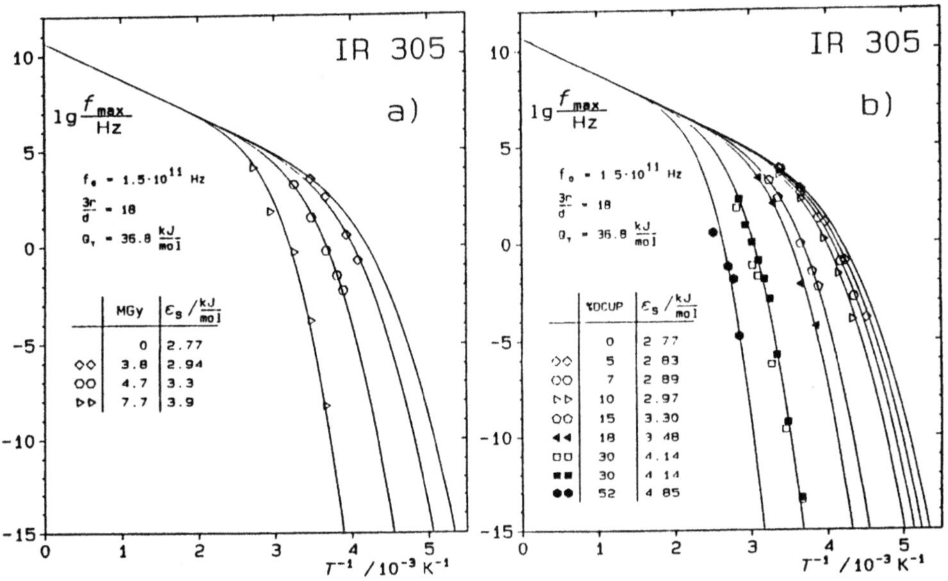

Fig. 17: Activation diagrams of two IR-series crosslinked (a) by radiation and (b) by DCUP.

The shearband process

The low frequency process is much more sensitive to crosslinking. We propose to attribute it to shearbands, if they are able to rearrange, thereby reducing their restoring forces [7]. The main experimental results are the dependences on the dose D or crosslinking density: $\Delta J_B^o \propto D^{-1}$, $f_B^o \propto D^{3.3}$, as is evident from Figures 14 and 13, respectively.

In the literature, the low frequency process is attributed to the slip of temporary entanglements and/or the onset of movement of pendant chains in the imperfect network [9-14]. Since, in the meander model of the bulk, the intrameander shear fluctuations determine the plateau compliance (no entanglements are needed), the low frequency relaxation can only be caused by dangling ends or mobile strands during shearband deformation: if a fraction $(1-\xi)$ of polymer segments has lost its orientation correlation (x_2-x_1) within the shearband, its relaxation strength should be increased from ΔJ_B^∞ to $\Delta J_B^\infty/\xi$ according to Equation (10) with $\Phi=1$. Such rearrangement is envisaged in Figure 7b, which might suggest that it will be a cooperative process of local tight fold reptation, provided its motion is not hindered by crosslinks.

----

[7] At low crosslinking there is still another more rigid component ΔJ_{BE}^o, f_{BE}^o, which we attribute to the shearbands when the chain ends only have rearranged. It can be detected because it is much faster than the main component but still separated from the glass process.

To put it quantitatively, let us assume that only that part l of a tight fold length L^+ [8], which is "open", may contribute to the rearrangement and lose its orientation correlation (cf. Fig.18 ,inset). Then, ξ becomes approximately

$$\xi \approx 1 - \frac{l_o}{L^+} \int_0^{L^+/l_o} \exp\left(-\frac{l}{l_o} p_c\right) d\left(\frac{l}{l_o}\right) \qquad (22)$$

(l_o monomer length, L^+ average tight fold length[8], p_c total crosslinking probability per monomer (=segment in the case of IR), and $\exp(-p_c l/l_o)$ the Poisson probability for l being free of crosslinks). In Fig. 18 $\Delta J_B^o/\Delta J_B^\infty = 1/\xi$ is double - logarithmically plotted versus p_c for three L^+/l_o values. It can be quantitatively compared with Figs. 14a,b, assuming the parameters: s/d=4/6, r/d≈7, $d^3/kT_o = 0.058$ MPa^{-1}, $\gamma_m = 9$, $\beta \approx 0.02$ and therefore $\Delta J_B^\infty = 0.40$ MPa^{-1} according to Equation (10, with $\Phi = 1$, $\xi = 1$) and $\Delta J_{eN}^o = 2.9$ MPa^{-1} according to Equ. (8, with P=1), if we also draw the ratio $\Delta J_{eN}^o/\Delta J_B^\infty$ into Fig. 18. Figs. 18 and 14 confirm that we have achieved a principal explanation of the relaxation strenghts versus crosslinking.

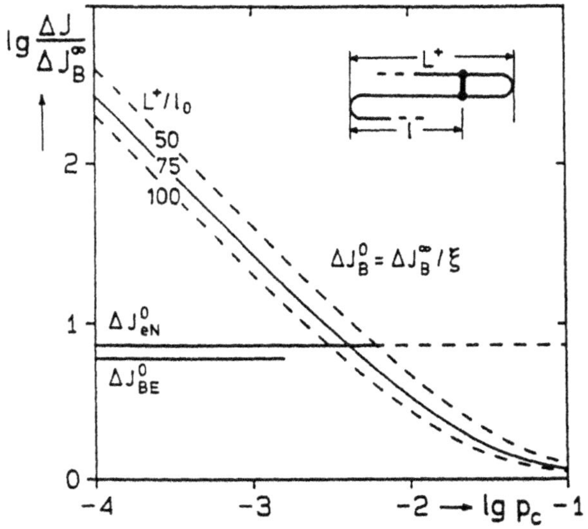

Fig. 18: Plot of the results of the model theory, to be compared with Fig.14a,b.

If we now compare e.g. the abscissa of the crossover point in Fig.18, say for the average tight fold lengths $1/p_f^o = L^+/l_o = 75$, i.e. $p_c = 4 \cdot 10^{-3}$, with those of the crossover points in Fig.14a,b, namely

a) $D_{crossover} \approx 0.166$ MGy
b) $DCUP_{crossover} \approx 0.41$ phr

we get the following relations between the total p_c and the amount of crosslinking agent

a) $p_c \approx 2.40 \cdot 10^{-2}$ D/MGy
b) $p_c \approx 0.97 \cdot 10^{-2}$ DCUP/phr

----

[8] "Tight fold length" L^+ is shorthand for the average chain length between tight folds.

A comparison of these p_c-relations with the previous ones (found from plateau compliance analysis for p_c^*) yields the same ratio as indicated in (21).

Moreover the ratio of effective crosslinks p_c^* for intrameander shear to the total crosslinking p_c turns out to be

$$p_c^*/p_c = 1/30 \qquad (23)$$

which is obout 1/4 of the value predicted by ideal contact probability consideration (19). p_c and therefore (23) rely on the model theory for ξ and the fold probability $p_f^o \approx 1/75$ assumed (from R_g-analysis [1]), and must be checked by equilibrium swelling. The appropriate theory in the frame of the meander model will be discussed in a subsequent paper.

6. The effect of filler on the dynamic shear compliance

In Fig.19 compliance master curves on uncrosslinked IR 305 rubber filled with different amounts (volume fraction Φ_F) of carbon black[9)] are presented. Its main effect is a strong shift of the whole curve of the matrix towards lower compliance with increasing filler content.

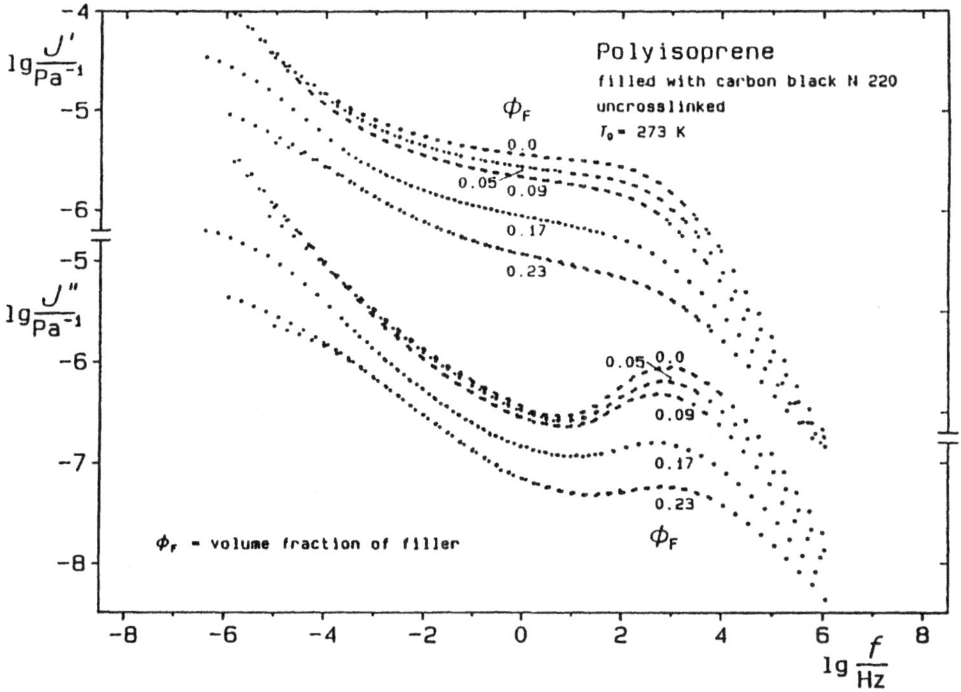

Fig. 19: Shear compliance master curves on uncrosslinked IR 305 rubber filled with different amounts of carbon black[9)]

The frequency of the glass relaxation stays about constant. The curve changes only little at low frequencies, i.e. at the onset of the flow region, where crosslinking would be strongly effective[10]. Therefore one has to conclude, that the filler particles - at least at low stress - do not primarily act as supercrosslinks, but shield part of the matrix from its normal deformation. To evaluate the measurements of Fig. 19, and those on slightly crosslinked samples[10], it is advantageous to plot the reduced compliance $J(\Phi_F)/J(0)$ versus filler content Φ_F (Fig.20).

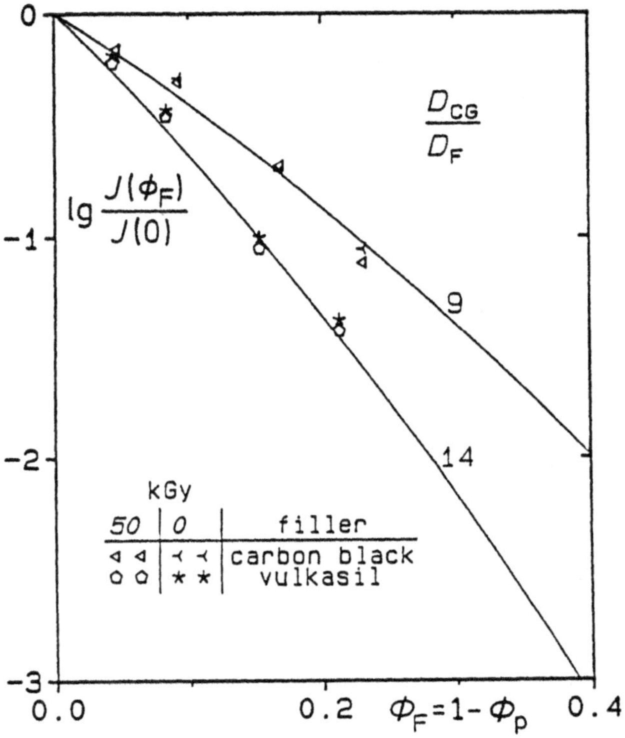

Fig. 20: The reduced shear compliance plotted versus the volume fraction of carbon black ($D_F \approx 30$nm) and precipitated silicate ($D_F \approx 20$nm)

Quantitatively, the filler effect on the zero shear compliance can be described on the level of the meander model and its coarse grain structure (Fig.21):
The paraelasticity of a file of meanders across one coarse grain of diameter D_{CG} is blocked up, if it contains at least a filler particle (of diameter D_F). This holds true for the intrameander shear (i.e. the plateau compliance) as well as for the shearband deformation.
According to this assumption shear deformation will only take place in those meander files not having embeded any filler particle, i.e. with the probability

$$J(\Phi_F)/J(0) = (1-\Phi_F)^{D_{CG}/D_F} \tag{24}$$

---

9) Carbon black, Corax N220 (Degussa), particle density 1.8

10) Measurements on similar samples which were radiation crosslinked (50kGy) after preparation show the shearband process (as in Fig.12a), which does not change its shape and relaxation frequency with filler content.

The curves in Fig.20 were drawn according to the relation (24) with D_{CG}/D_F as a fit parameter. From the numbers found for best fit, one obtains for the average diameter of the coarse grains in isoprene rubber

$D_{CG} \approx 9D_F$ carbon black
$\approx 14D_F$ silicate
≈ 300 nm

because the diameters of the filler particles are roughly known to be 30 and 20nm, respectively.

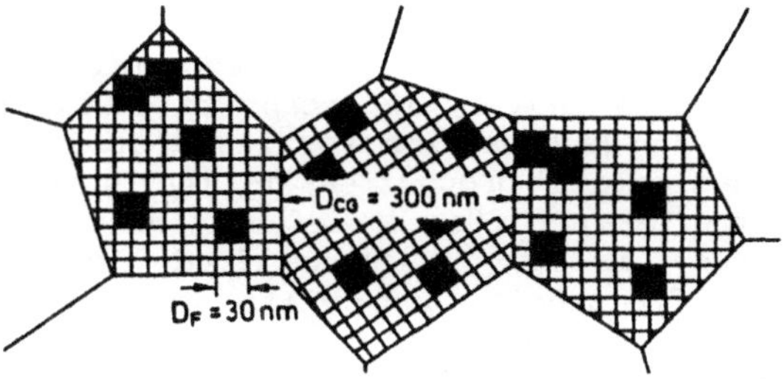

Fig. 21: The coarse grain structure of the polymer matrix with embeded filler particles blocking the shear deformation of several meander files within the grains.

The predicted particle size dependence of the zero shear compliance reduction (24) requires further check, if possible on a series of particles of similar nature but differing in diameter. The first step towards this aim is shown in Fig.22 on 3 60 phr carbon black filled samples, using N330, N220, N110 as filler. Their diameters are not yet known with sufficient accuracy to try a quantitative check of (24), but the compliance reduction is in the right direction.

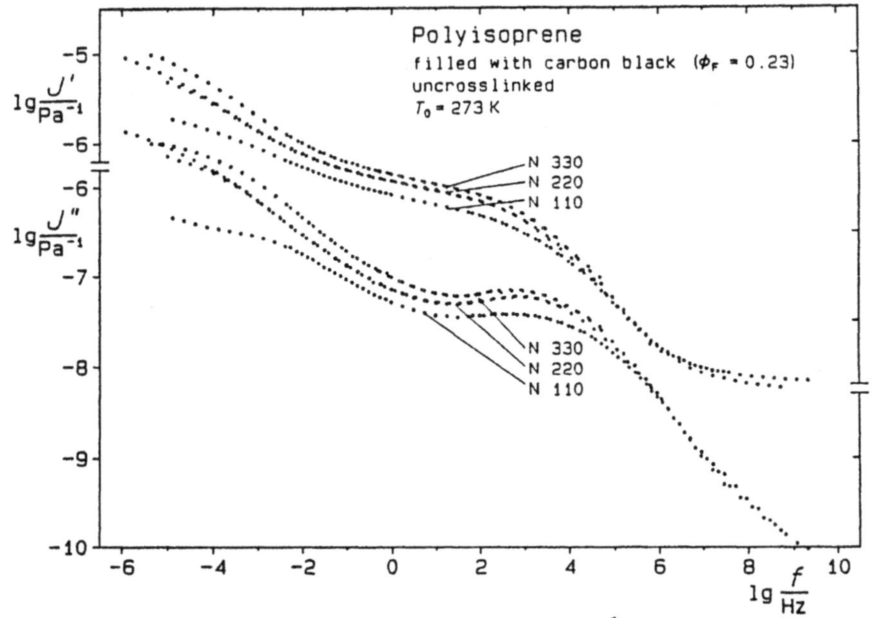

Fig. 22: Shear compliance of 3 samples, filled 60 phr with carbon black particles of different size (N330>N220>N110).

7. Conclusion

The study of the dynamic mechanical compliance of polymer melts and networks is very important to check molecular models in detail. This is a presupposition for its application to nonlinear phenomena, e.g. rheological properties or the stress-strain behaviour up to large strains in different deformation modes. The next step, which is under study, is that of the amplitude dependence of the relaxation strengths and frequencies and/or its dependence on superimposed stress, strain or shear rate.

Acknowledgement
The authors wish to thank Dr. U. Eisele for supplying the samples, the Deutsche Forschungsgemeinschaft (SFB 239) and the Fonds der Chemie for financial support, and Mrs. Auer-Oppitz and Mrs. Schiffner-Großmann for preparing the manuscript.

References

[1] Pechhold W, Böhm M, v.Soden W (1987) Coll&Polym Sci 75:23
[2] v.Soden W (1988) Thesis,Ulm
[3] Pechhold W, v.Soden W, Stoll B (1981) Macromol Chem 182:573
[4] Pechhold W (1984) Macromol Chem, Suppl 6:163
[5] Pechhold W, Gross T, Großmann HP (1982), Coll&Polym Sci 260:378
[6] Pechhold W, Großmann HP (1979) Faraday Discuss Chem Soc 68:58
[7] Pechhold W, Sautter E, v.Soden W, Stoll B, Großmann HP (1979) Macromol Chem, Suppl 3:247
[8] Pechhold W (1980) Coll&Polym Sci 258:269
[9] Havránek A, Ilavský M, Nedbal J, Böhm M, v.Soden W, Stoll B, (1987) Coll&Polym Sci 265:8
[10] Ferry JD (1980) Viscoelastic Properties of Polymers, 3rd ed., Wiley, New York
[11] Böhm M, v.Soden W, Heinrich W, and Yehia AA (1987) Coll&Polym Sci 265:295
[12] Valles EM, Macosko CM (1979) Macromolecules 12:673
[13] Havránek A (1982) Polym Bull 8:133
[14] Bibbo MA, Valles EM (1984) Macromolecules 17:360
[15] Heinze HD, Schmieder K, Schnell G, Wolf KA (1961) Kautschuk und Gummi 14:208 WT

PARAELASTICITY OF ORGANIC COLLOID SYSTEMS

W. PECHHOLD
Abteilung Angewandte Physik, Universität Ulm
Oberer Eselsberg
7900 Ulm, F.R.G.

ABSTRACT: The structure and the thermodynamic and electric properties of dense, organic colloid systems should be studied in conjunction with their mechanical properties to understand the role of interfaces and to have a further check on the structural models assumed. Such study can be carried out experimentally and by model theories. In this paper the dynamic shear compliance of two anisotropic colloids is reported (for a clay organic complex and an oriented stack of DPPC-lamellae) and interpreted by dislocation-mediated shear across the interfaces. An additional relaxation process in DPPC is attributed to a switching of the tilted unit cells between equivalent states. Moreover, the shear compliance in the "rotor phase" of the n-alkanes and the plateau-compliance of polymer melts can be described by the combined shear modes of cubic blocks, the size of which is determined by the molecule length (in the paraffins) or by the meander size (meander theory for polymer melts). It is worth mentioning that a single formula can explain to a first approximation all of the four examples reported.

1. Introduction

In dense colloid systems the interfaces between the particles (e.g. domains, grains, lamellae etc.) determine the mechanical properties. This is accomplished mainly in two ways: (i) The misfit between the particles allows the particles to deform or to interchange between equivalent states without increasing the energy. (ii) The interfaces easily permit the creation and annihilation of dislocation loops, mediating additional shear-defomation or even viscous flow between the particles.

In the following the paraelastic behaviour of two anisotropic colloids (a clay-organic mineral and an oriented stack of DPPC-lamellae) and two isotropic colloids (paraffin wax and polymer melt) will be discussed.

O. Güven (ed.), Crosslinking and Scission in Polymers, 223–237.

2. Paraelastic relaxation strength

To derive a formula for the relaxation strength of domains (kink-blocks, tilted chain-blocks), switching between two equivalent states (fig.1), one may simply consider the free energy of shear deformation,

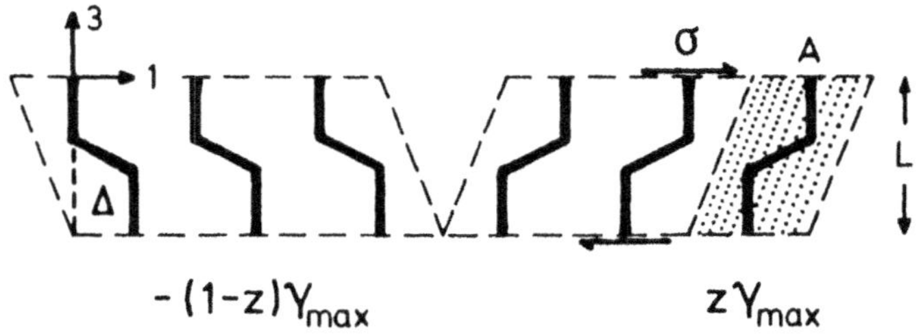

Fig. 1 Two-state model of kink-block orientation

$$\frac{\Delta g_{def}}{kT} = z\ln z + (1-z)\ln(1-z) + \frac{v\sigma}{kT}[\gamma-(2z-1)\gamma_{max}] \tag{1}$$

asking for the equilibrium z (and hence γ) via minimization. The first two terms comprise the ideal mixing entropy of both states, the third term represents the necessary condition for the total shear γ, with $v\sigma/kT$ as the Lagrangian multiplier. The latter relation follows from the definition $\partial(\Delta g/v)/\partial\gamma = \sigma$, v being the volume of the statistical element (e.g. a kink in a small kink-block). Minimization yields

$$\gamma = \gamma_{max} \cdot \tanh(v\sigma\gamma_{max}/kT) \approx v\sigma\gamma_{max}^2/kT \quad (\sigma\to 0) \tag{2}$$

and the relaxation strength becomes (with $v = A\cdot L$, $\gamma_{max} = \Delta/L$)

$$2\Delta s_{1313} \equiv \partial\gamma/\partial\sigma \approx \frac{v}{kT}\gamma_{max}^2 = \frac{A\cdot L}{kT}\left(\frac{\Delta}{L}\right)^2 = \frac{N}{V}\frac{A^2\Delta^2}{kT} \tag{3}$$

where $1/v = N/V$ is the number of elements per unit volume.

Fig. 2

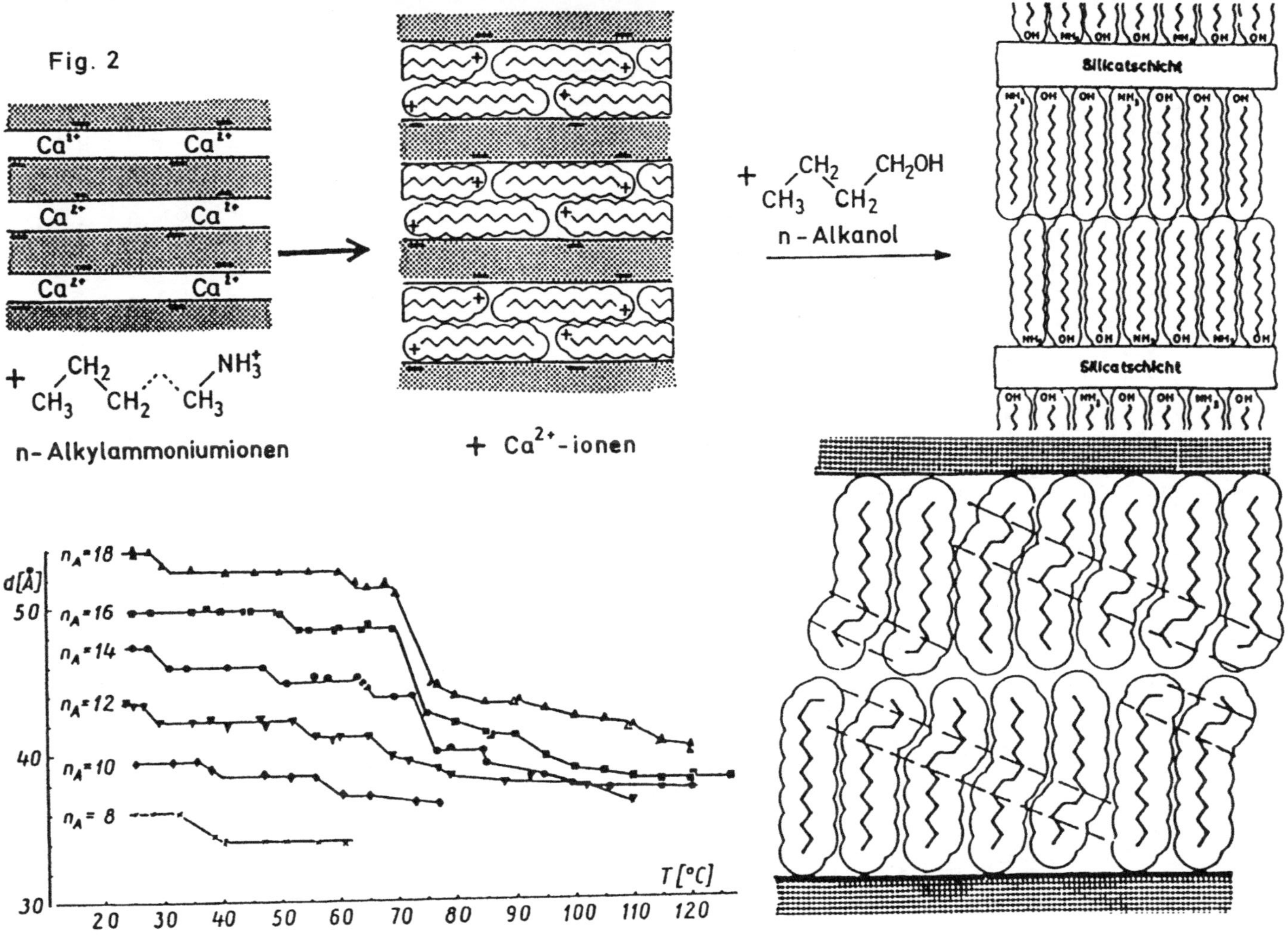

Silicatschicht
Silicatschicht
+ $CH_3CH_2CH_2CH_2OH$
n - Alkanol
+ $CH_3CH_2CH_2 \cdots CH_2NH_3^+$
n- Alkylammoniumionen
+ Ca^{2+}-ionen
d[Å]
T[°C]
n_A=18
n_A=16
n_A=14
n_A=12
n_A=10
n_A= 8

3. Clay organic complexes

Lagaly and *Weiss* have prepared n-alkylammonium-n-alkanol-beidellite- or vermiculite-complexes by cation exchange in Ca^{2+}-beidellite or vermiculite with n-alkylammonium, and subsequent drying and swelling in n-alkanol. They have described the organic bilayer-structures formed, and carefully measured their basal spacings from the melting point of the n-alkanol to higher temperatures. In fig.2 the preparation of the model systems and the results got from the X-ray-diffraction are summarized as taken from [1,4]. With rising temperature the basal spacing decreases in distinct steps of 1.2 ± 0.1Å up to a transition temperature from the β into the α-phase (indicated by a larger step of 4-6 Å [2,3,4]).
Lagaly and *Weiss* explained these steps as the consequence of cooperatively formed kink-blocks [1,2,3,4] and *Pechhold et al* [5] confirmed this view by a statistical model theory applying Cluster-Entropy-Hypothesis (CEH). It should be kept in mind that there is a thermodynamic driving force towards kink-block formation in these clay organic complexes: the available cross-section per all-trans chain is 23-25 $Å^2$ (due to the pockets in the pseudohexagonal silicate layers) compared with that of the best intermolecular fit (18.2 $Å^2$ in PE).

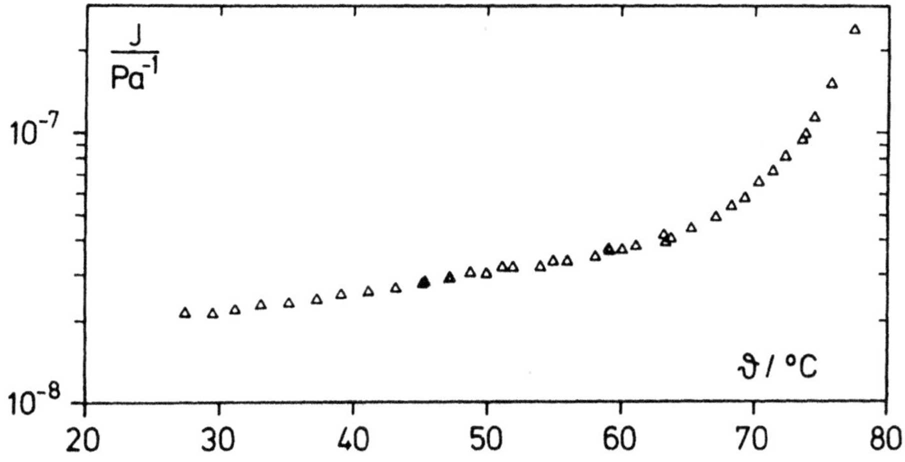

Fig. 3 Real part of the shear compliance of n-dodecylammonium-n-dodecanol-vermiculite at 12 kHz

Mechanically these clay organic complexes behave like wax parallel to the basal planes: at low frequencies they show a visco-plastic behaviour, but in the kHz-range they are elastic as reported in fig.3: the real part J' of the shear compliance (by a factor of 10 larger than the loss-compliance J") is smoothly depending on temperature with increasing slope only near to the β/α-transition. A first check of the kink-block model discussed above requires a comparison of the (nearly, i.e. within the present experimental error) missing J-steps (at 28° and 55°C) with the prediction from model theory. Applying formula (3) with an additional factor of 1/2 (assuming half of the unit cells to contribute in one shear direction) one may use the following data

$\Delta^2 = (\delta_y^2+\delta_z^2) \approx 4.73$ Å^2 from ideal lateral kink displacements δ_y, δ_z [6]

$T \approx 320$ K,

$A = 2\cdot 24.5$ Å^2 (two pairs of molecules/unit cell), $L \approx 43$ Å,

and gets $\Delta J_1 = 2\Delta s_{1313} \approx 6.1\cdot 10^{-10}$ m^2/N for the single unit cell as the statistical element. More precise measurements of the shear compliance simultaneous with the sample thickness of perfectly swollen complexes will test this relaxation strength prediction. The view that the whole kink-blocks might act as statistical elements would lead to $\Delta J_b \approx \Delta J_1 \cdot L/2c \approx 5\cdot 10^{-9}$ m^2/N and can be excluded by the experimental result (fig.3), which therefore may be considered as a further proof of CEH.

The second question concerns the absolute value of the shear compliance and its smooth temperature dependence (fig.3). Since a shear compliance of $2\ldots 5\cdot 10^{-8}$ m^2/N cannot be explained by energy elasticity, it is supposed to be due to a paraelastic shear in the central interface of the bilayers. The model proposed here (fig.4) assumes creation and annihilation of dislocation loops across these interfaces. The loop density (or size) in equilibrium can be estimated by free energy consideration (applying CEH), thereby taking into account

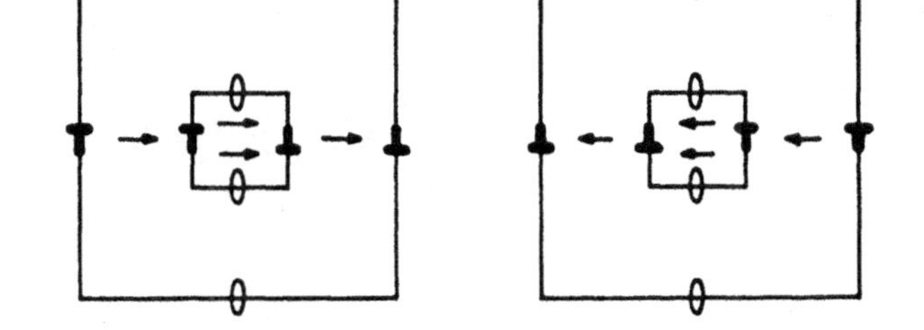

Fig. 4

Dislocation-loop model to explain interface paraelasticity of less dense organic lamellar systems

dislocation self-energy and interaction as well as the relation between the total shear and the dislocation density of both signs. This model theory which also explains the temperature dependence of the shear compliance will be published elsewhere. It is worth noting that the elastic modulus included in the dislocation energy is due to chain bending. This makes it clear why such interface paraelasticity can only occur in lamellae with less dense packing than in the ideal chain crystal. To have a rough estimation of the shear compliance here one may assume average slip areas $A \approx (L/2)^2$ to exist in three states and produce local shears of d/L, 0, -d/L respectively. Then one can apply formula (3) (with an additional factor of 2/3 which follows from a similar derivation). Using the data

$$\Delta \equiv d = \sqrt{24.5}\ \text{Å} \quad \text{for the average chain distance}$$

$$A = (L/2)^2, \quad L = 43\ \text{Å} \quad \text{and} \quad T \approx 320\ \text{K}$$

one gets $J = 2\Delta s_{1313} \approx 4 \cdot 10^{-8}\ m^2/N$, a value not too far from the experimental result (but with a wrong dependence on temperature). The increasing slope of the compliance starting at 65°C indicates the onset of larger shear fluctuations which finally lead to tilted chain domains and a larger step in the basal spacing.

4. Oriented stacks of lipid lamellae

Dipalmitoyl-Phosphatidylcholine (DPPC) was incorporated with water at 90° to 120°C (i.e. in the LC-phase) in the vapour pressure of K-acetate/water solutions, pressed to about 0.2 mm thickness between silanated glass slides, and after 4 hours annealing slowly cooled to room temperature [7]. Fig.5 shows a freeze-fractured and Pt/C-shadowed electron-micrograph, almost perpendicular to the multilayers [7]. Such samples of different water content were glued in pairs into the double sandwich holder of the mechanical spectrometer and measured in the frequency range 10^{-4} to 200 Hz at different temperatures (30° to -30°C, i.e. in the $L_{\beta'}$ - or gel-phase). The water content was kept about constant by appropriate vapour pressure in the spectrometer cell. Quantitatively the water content was

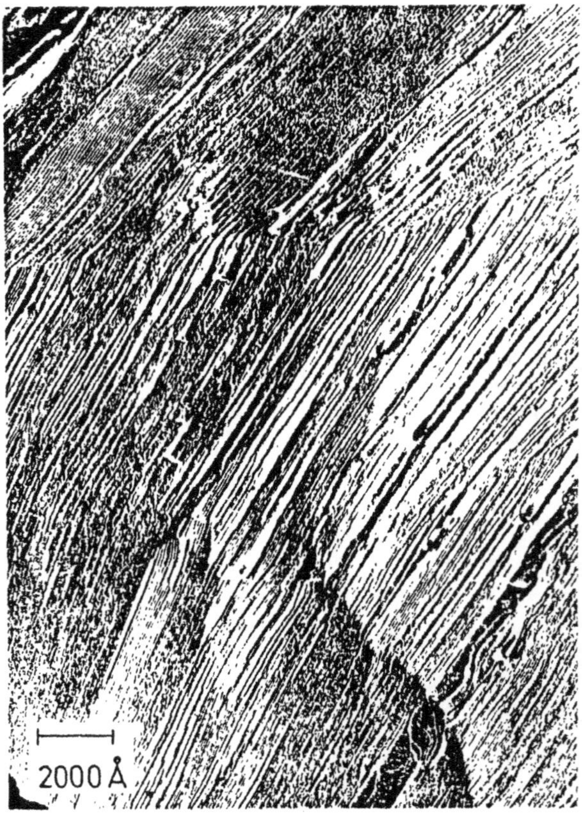

Fig.5 Freeze-fracture electron-micrograph of oriented DPPC

determined on excess samples before and after the measurement by coulombmetric titration using the Karl-Fischer-methode (Mitsubishi, moisture meter, model CA-02).

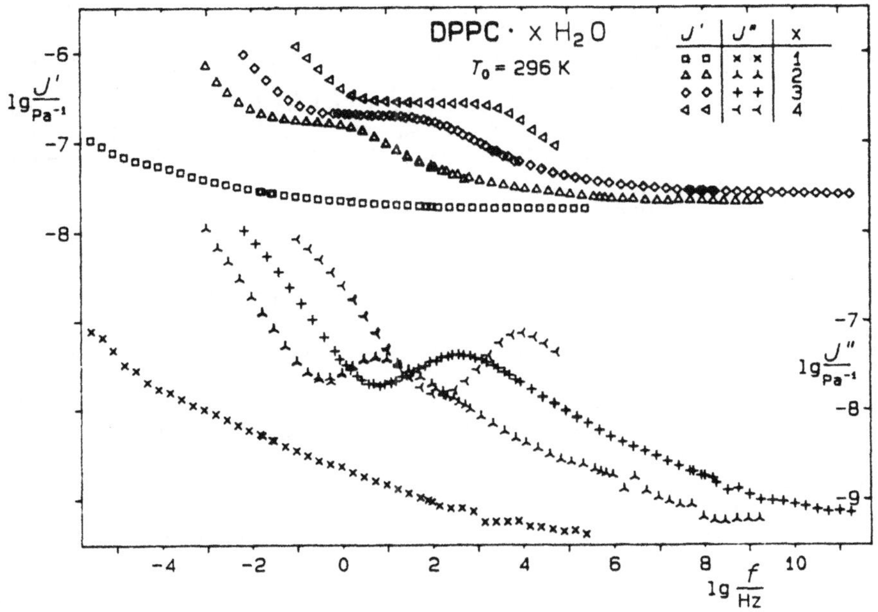

fig. 6 Shear compliance mastercurves of oriented DPPC in the $L_{\beta'}$-phase reduced to 296 K at different water contents

The frequeny-curves measured on each sample at different temperatures could be combined to give a mastercurve (T_0=296 K). Four of them are shown in fig.6 and will now be discusssed. (i) The high frequency plateau of J' is about $2\cdot10^{-8}$ m^2/N, a value similar to that of the clay organic complexes. Therefore its interpretation probably must be the same, i.e. a dislocation-loop mechanism within the hydrophobic interface of the lecithin-bilayer. (ii) At very low frequencies the onset of flow relaxation and viscous flow can nicely be seen. It seems to be characteristic for the waxy state as well as for the polymer melt, eventhough the retarding mechanisms will be different.
(iii) In the intermediate frequency range a strongly developed relaxation process takes place, which probably is due to a switching of the tilted unit cells between energetically equivalent states (in differently tilted domains).

Fig.7 gives a schematic drawing (according to the structure analysis in [8]) of two such domains tilted by an angle Θ. Applying formula (3) with an additional factor of 1/2 (assuming half of the unit cells to switch in one shear direction) one gets for the relaxation strength

$$\Delta J \equiv 2\Delta s_{1313} \approx \gamma^2_{max} \cdot A \cdot L/2kT \qquad (3a)$$

Fig. 7

Schematic drawing of the DPPC-bilayer with oppositely tilted domains

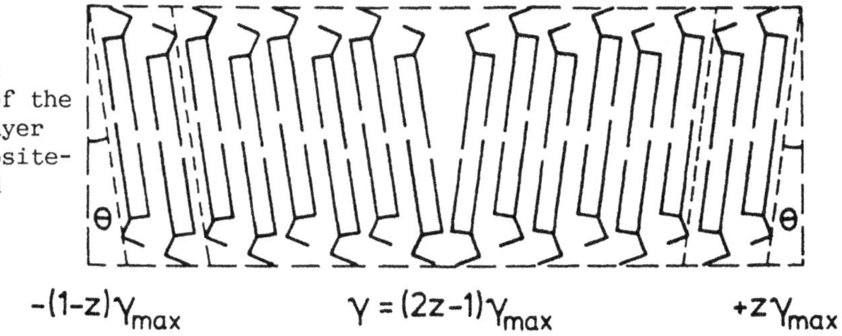

In fig.8a there are plottet the tilt angles Θ = arctan γ_{max} calculated from the experimental relaxation strengths, using the data

$$L = 57 \text{ Å}, \quad A = 4 \cdot 39 \text{ Å}^2 \text{ (4 pairs of molecules/unit cell)}, \quad T = 296 \text{ K}$$

This tilt-angle Θ includes the tilting of hydrocarbon chains Θ_h as well as the tilting of the polar heads Θ_p. In the gelphase ($L_{\beta'}$) $\Theta_h \approx 12 \cdot\cdot 13°$ obviously does not depend on water content [8,9,10]. Since the lamellar repeat L (fig.8b) stays about constant between the di- and tetrahydrate it must be assumed that Θ_p as well as Θ increases. The latter result is also shown by the tilt angles Θ deduced from the mechanical relaxation strengths via (3a).

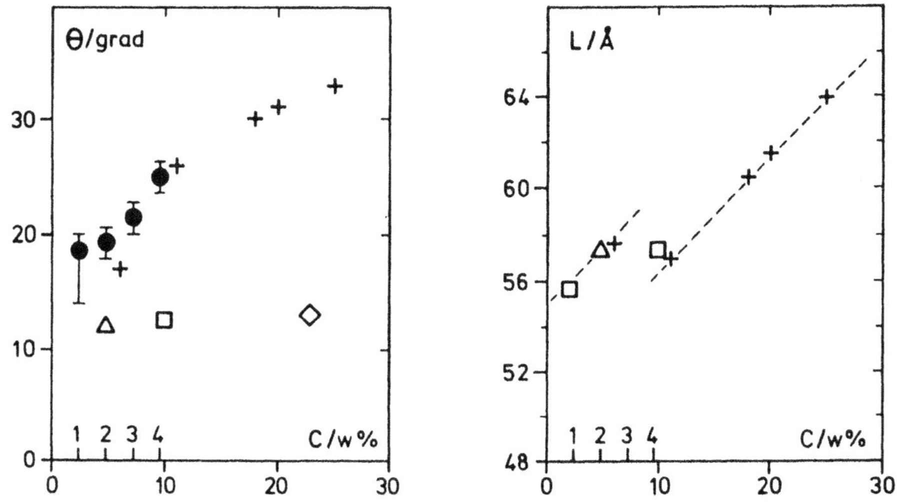

fig. 8 (a) tilt angle Θ_h of the hydrocarbon chains in DPPC, Δ[8], □[9], ◇[10]. Overall tilt angle Θ from X-ray and specific volume +[11] and from paraelastic relaxation strength ●[this work]

(b) Lamellar repeat L in DPPC in dependence on water content (from the same sources)

5. The n-alkanes in the waxy state

The n-alkanes exhibit up to four (A-D) modifications below ultimate melting. A detailed description of the molecular packing (lateral and end-to-end) and motion in these phases has been given for $C_{33}H_{68}$ [12] mainly from X-ray, Raman, dielectric, NMR and INS analysis.

Dynamic mechanical measurements at 12 kHz [13] (cp. fig.9) show, that only the quasihexagonal modification (the so-called "rotor phase" below the melting point) exhibits strong paraelasticity (and waxy flow at low frequencies). Its interpretation in [13] - assuming switching of highly concentrated kink-blocks (without applying CEH at those days) - was disproofed by *Strobl's* finding [12] of only 0.4-0.7 kinks per chain in the D-phase. No further explanation for the paraelastic plateau in the kHz-range was put forward since that time.

Taking *Strobl's* analysis of the D-phase, the following model can now be suggested for its paraelasticity (fig.10): assuming a cubic block of chains as the statistical element for shear deformation of the lamellae, there will be three contributions of anisotropic relaxation strength.

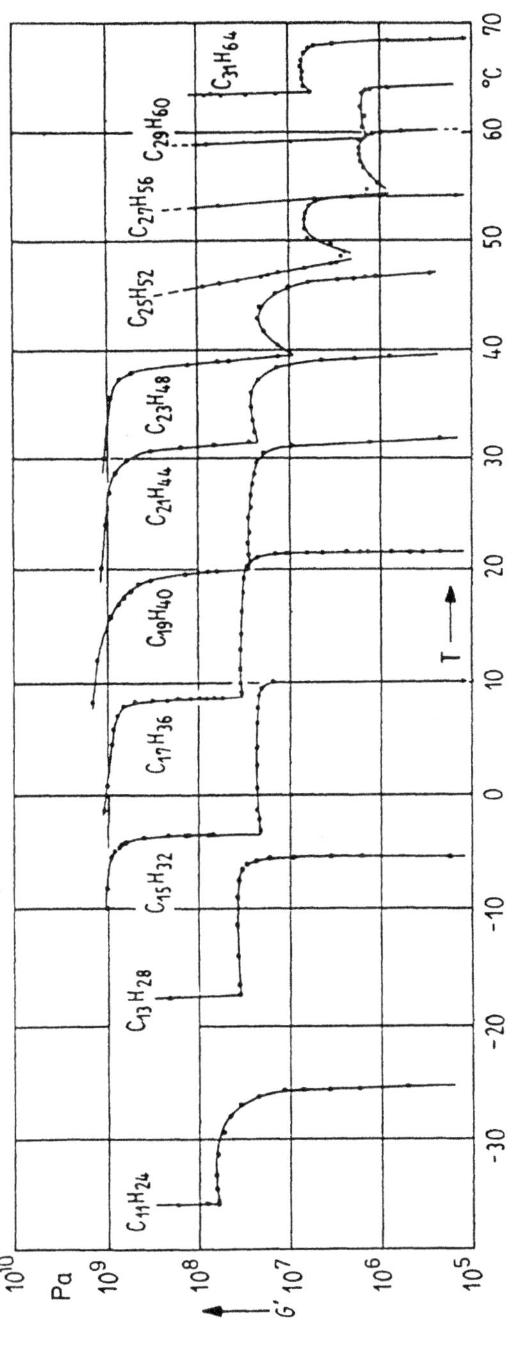

Fig. 9

The real part G' of the shear modulus at 12 kHz of isotropic n-alkanes near to the melting point. (G" is some 5 to 10 times smaller than G').

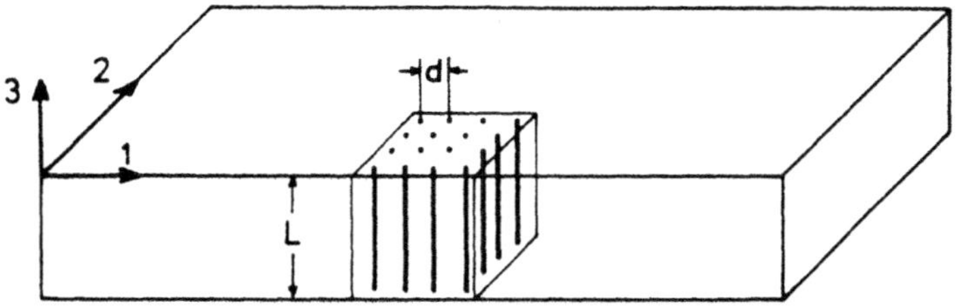

Fig. 10 A cubic block of chains as the statistical element for paraelastic shear in the "rotor phase" of n-alkanes

(i) Two contributions of interlamellar shear

$$2\Delta s^{(1)}_{1313} = 2\Delta s^{(1)}_{2323} = (d/2L)^2 \cdot L^3/kT \qquad (3b)$$

due to the dislocation loop mechanism, with two states of local shear ($\pm d/2L$) assumed, and

$$2\Delta s^{(2)}_{1313} = 2\Delta s^{(2)}_{2323} = (\tan\Theta)^2 A \cdot L/2kT \qquad (3c)$$

due to the switching of tilted chains, assuming half of the unit cells to interchange in one shear direction.

(ii) a paraelastic shear of the cube in direction of the n-alkane chains, consisting of multiples of c/2 glide steps, which every row of molecules may undergo relativ to its neighbour

$$2\Delta s_{3131} = 2\Delta s_{3232} = \frac{1}{3}\, m(m+1)(c/2d_\perp)^2 d_\perp L^2/kT \qquad (3d)$$

This formula follows from a derivation similar to that of (3).
m, the maximum number of steps to each side, is restricted by one half of the overall thickness of the void layers between the lamellae. This has been determined with SAXS by *Strobl* [12] to be ~10/2 Å for $C_{33}H_{68}$. Therefore m=2 is most probable.
To calculate the isotropic shear compliance one favourably applies Reuss-averaging, which in this case reads

$$\Delta J_{iso} = \frac{1}{5}\,(\Delta s_{1313} + \Delta s_{2323} + \Delta s_{3131} + \Delta s_{3232}) \qquad (4)$$

Using the data [12]

$A = 2 \cdot 20.8$ Å^2, $d = 4.9$ Å (chain distance), $d_\perp = d\sqrt{3}/2 = 4.25$
L/Å = 1.27 n (length of trans-chain, n number of C-atoms)
c = 2.54 Å (identity period), $\Theta = 19.5°$ (tilt angle), m = 2,

the formulae (3b-d, 4)yield the isotropic shear compliances/moduli in dependence of the chain length which are listed in table 1. The respective temperatures were taken from fig.9.

n-alkane	T_D/K	$2\Delta s_{1313}^{(1)}$	$2\Delta s_{1313}^{(2)}$	$2\Delta s_{3131}$	$\Delta J\cdot 10^7/\frac{m^2}{N}$	$\approx G'\cdot 10^{-7}/\frac{N}{m^2}$
$C_{11}H_{24}$	240	0.253	0.110	0.447	0.162	6.17
$C_{15}H_{32}$	275	0.301	0.131	0.726	0.232	4.32
$C_{19}H_{40}$	300	0.350	0.152	1.067	0.314	3.19
$C_{23}H_{48}$	315	0.403	0.175	1.49	0.414	2.42
$C_{29}H_{60}$	335	0.478	0.208	2.23	0.583	1.72
$C_{33}H_{68}$	343	0.531	0.231	2.82	0.716	1.40

table 1: Paraelastic shear compliance/modulus relaxation strength in the "rotor phase" calculated from model theory in dependence on chain length (cp. text)

The agreement of theoretical prediction (table 1) and experiment (fig. 9) is quite good, including $C_{33}H_{68}$ which has been measured separately at 1.2 kHz (fig.11), therefore G" is much higher than at 12 kHz. A large deviation is observed in the chain length range $C_{25}\cdots C_{31}$ (fig. 9) which one might speculate may be due to m=4 c/2-glide steps in each chain direction, combined with a macroscopic orientation (the n-alkanes arranging parallel to the surface of the thin sample).

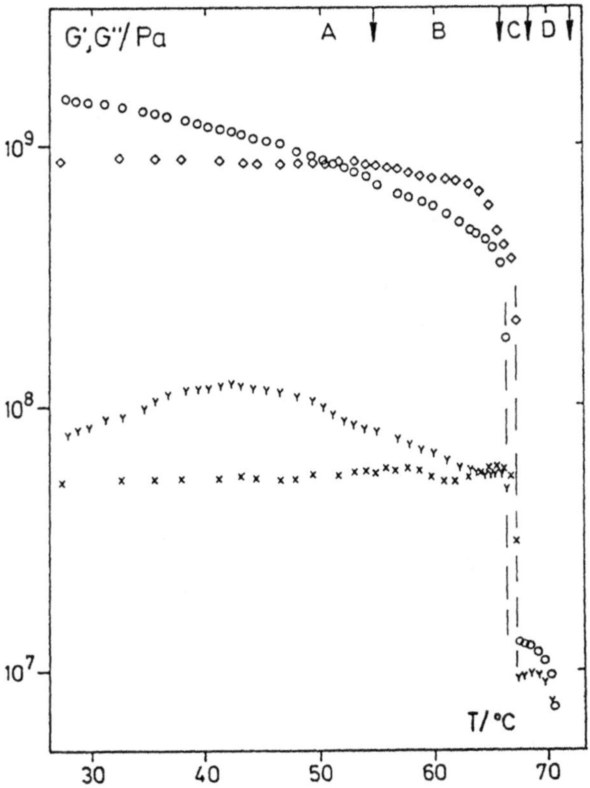

fig. 11 Dynamic shear modulus of $C_{33}H_{68}$ measured at 1.2 kHz ◇× pressed powder, crystallized from petrol ether dilute solution ; ○Y after heating into the waxy state

6. Polymer melts and networks in the rubbery state (plateau compliance)

The most remarkable property of polymer melts or networks is rubber elasticity, which can only be explained by assuming paraelastic segmental movement yielding entropic restoring forces. This can be nicely studied by analyzing dynamic shear compliances versus frequency. For melts and amorphous networks, the frequency/temperature superposition is valid, as a rule, yielding mastercurves (over more than 20 decades in frequency, fig.12) and the proper shift factors on both axes.

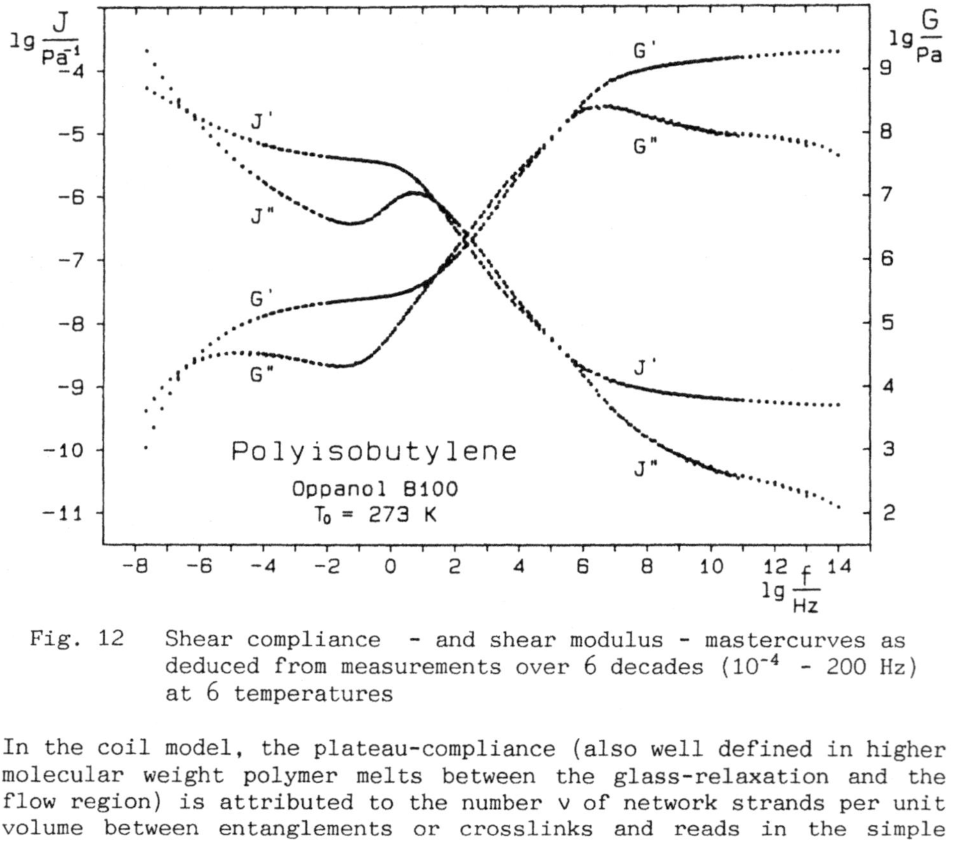

Fig. 12 Shear compliance - and shear modulus - mastercurves as deduced from measurements over 6 decades (10^{-4} - 200 Hz) at 6 temperatures

In the coil model, the plateau-compliance (also well defined in higher molecular weight polymer melts between the glass-relaxation and the flow region) is attributed to the number ν of network strands per unit volume between entanglements or crosslinks and reads in the simple case

$$J_{eN} = (R_0^2/R^2)/\nu kT \qquad (5)$$

In the meander model, the plateau compliance can be explained by the relaxation-strengths of intra-meander shear deformation modes (fig.13) applying Reuss-averaging. In the statistical derivation of the bulk anisotropic compliances, layers of molecules are allowed to be displaced by one chain distance d against each other. This theoretical model has been worked through in detail in preceding papers [14,15,16] and yields, for the plateau compliance J^0_{eN} of an uncrosslinked polymer melt

$$J^0_{eN} = \frac{1}{5}\left\{\begin{matrix} s_{1212} + s_{2121} \\ + s_{3131} + s_{3232} \end{matrix}\right\} = \frac{d(r+x)^2}{9kT} \qquad (6)$$

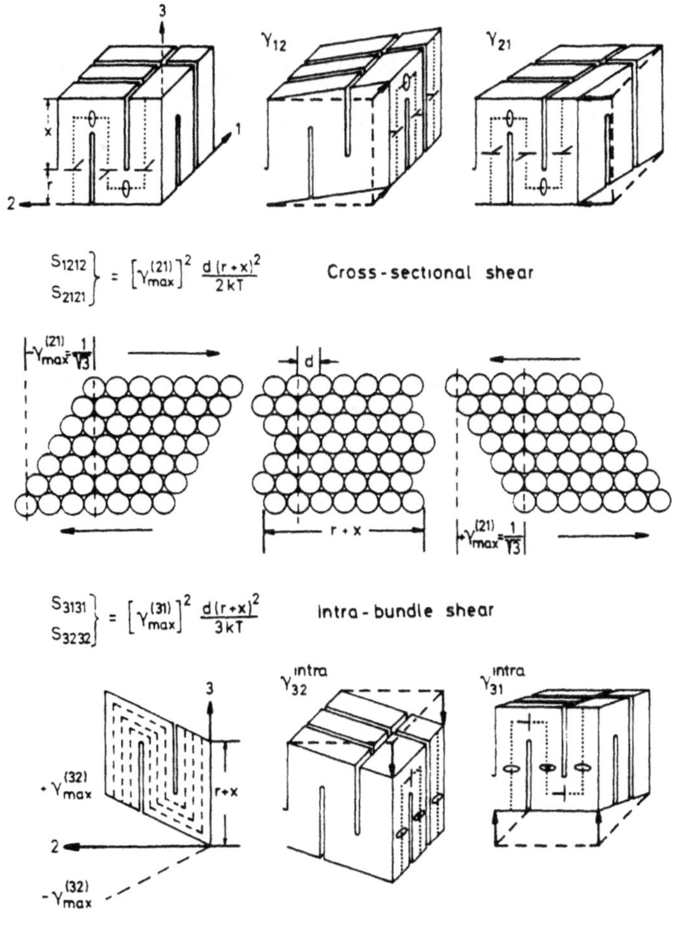

Fig. 13 The two main intra-meander shear modes demonstrated on single meander cubes with dislocations indicated on the surfaces

Taking J^0_{eN} from the decomposition of master curves of various polymer melts, equation (6) represents probably the best method to determine the cube side length r+x or the average number of quasiparallel stems (r+x)/d in a layer of molecules [17].

Conclusion

The study and interpretation of paraelastic properties of organic colloid systems is a powerful tool to understand the role of interfaces and to have a further check on the structural models assumed. The examples presented here shall be looked at as a first attempt towards this aim. The similarity in the approach to the "rotor phase" of the n-alkanes (though still crystalline) with that to the polymer melts and networks gives additional support to the meander model of amorphous polymers. (The rotation of the meander cubes across their space diagonals does not contribute to the plateau compliance.) The tilt-switch relaxation in the DPPC lamellae could be the important mechanism driving the gate-proteins in biological membranes. Further examples of paraelasticity to be dealt with in the future will include the deformation and swelling behaviour of biopolymers and biological structures e.g. the cornea and sclera of the eye.

Acknowledgement

The authors wish to thank the Deutsche Forschungsgemeinschaft and the Fonds der Chemie for financial support.

References

[1] Lagaly G., Fitz S., Weiss A. (1975) Prog.Coll.&Polym.Sci. 57:54

[2] Lagaly G., Fitz S., Weiss A. (1975) Clays and Clay Minerals 23:45

[3] Lagaly G. (1976) Angew. Chemie 88:628

[4] Lagaly G., Weiss A. (1971) Koll.Z.u.Z.Polymere 248:979

[5] Pechhold W., Liska E., Grossmann H.P., Hägele P.C. (1976) Pure & Appl.Chem. 46:127

[6] Pechhold W. (1968) Koll.Z.u.Z.Polymere 228:1

[7] Jürgens E. (1981) Thesis Universität Ulm

[8] Hauser H., Pascher I., Pearson R.H. and Sundell S. (1981) Biochim.Biophys.Acta 650:21

[9] Stamatoff J.B., Graddick F., Powers L. and Moncton D.E. (1979) Biophys.J. 25:253

[10] Hentschel M., Hosemann R. and Helfrich W. (1980) Z.Naturforschung 35a:643

[11] Tardieu A., Luzzati V. and Reman F.C. (1973) J.Mol.Biol. 75:711

[12] Ewen B., Strobl G.R., Richter D. (1980) Farad.Disc.Chem.Soc. 69:19

[13] Pechhold W., Dollhopf W. and Engel A. (1966) Acustica 17:61

[14] Pechhold W.R. (1980) Coll.& Polym.Sci. 258:269

[15] Pechhold W. (1984) Makromol.Chem., Suppl. 6:163

[16] Pechhold W., Böhm M. and v.Soden W. (1987) Progr.Coll.& Poly. Sci. 44:23

[17] Pechhold W., Sautter E., v.Soden W., Stoll B., Großmann H.P. (1979) Makromol.Chem., Suppl. 3:247

THE USE OF SIZE EXCLUSION CHROMATOGRAPHY IN THE STUDIES OF CROSSLINKING AND CHAIN SCISSION IN POLYMERS

Olgun Güven

Department of Chemistry
Hacettepe University
Beytepe, Ankara, 06532, Turkey

INTRODUCTION

Depending on the various modes of initiation Crosslinking and Chain Scission processes may be affected by chemical agents, heat, light, high-energy radiation, shearing forces, ultrasonic vibrations, rapid freezing of a polymer solution, high speed stirring, enzymes, bacteria, etc. in natural and synthetic polymers. The two main processes discussed in this meeting, namely crosslinking and chain scission cause important changes in the size of polymer chains. When one is concerned with the changes produced in the size of polymer chains either due to degradation which leads to chain scission or crosslinking which causes an increase in thir sizes, SEC becomes one of the most useful techniques provided that the polymer under investigation retains its solubility in the solvent studied. Size Exclusion Chromatography (SEC) is an important technique for determining the molecular weight and molecular weight distribution of polymers in dilute solution.

O. Güven (ed.), Crosslinking and Scission in Polymers, 239–250.

CHAIN SCISSION

Depending on their chemical structures and presence or absence of oxygen, polymers can be classified either as mainly degrading or crosslinking type when subjected to various modes of initiation. SEC becomes the standart method of choice when the polymer studied is predominantly of degrading type. Although the degradation either in the form of main chain scission or unzipping can be followed by any method of polymer molecular weight determination, a typical SEC chromatogram provides not only the individual values of weight average M_w , number average M_n, or any other higher or lower average molecular weights but also molecular weight distribution of the sample studied.

The value of obtaining a whole molecular weight distribution curve by SEC is shown in the degradation of natural rubber during mastication (1). This study indicates that as the rubber is milled the molecular weight distribution becomes narrow which is probably due to mechanical degradation of large molecules. Depending on the milling time a high molecular weight tail is also observed indicative of concurrent crosslinking. The advantages of using SEC technique in the degradation of polystyrene by gamma irradiation are also shown in the work of Bowmer et al. (2). Scission and crosslinking yields are determined as a function of depth by SEC using primary calibration of layers removed progressively from the surface. It is determined that chain scission decreased and crosslinking increased with depth in accordance with increased scission by reaction with oxygen and a diffusion effect.

When polymers are irradiated with high-energy radiations, quantitative aspects of such a process can be analyzed after determining various molecular weight averages. For polymers which undergo random chain scission and crosslinking, the number average molecular weight of the exposed polymer, M_n, can be expressed in terms of the initial, $M_{n,o}$, the dose received, D, and the so-called scission and crosslinking yield values, G (S) and G(X) respectively as given in eqn.(1)

$$1/M_n \quad = \quad 1/M_{n,o} \; + \; [\, G(S) - G(X) \,]\, D \; / \; 100\, N_A \qquad (1)$$

where N_A is Avogadro' s number and the G values are defined as the number of events produced by the absorbtion of 100 ergs of radiation. The determination of radiation chemical yields, i.e. G values is possible through the use of eqn.(1) simoultaneously with eqn.(2).

$$1/M_w = 1/M_{w,o} + [G(S)/2 - 2G(X)]\ D / 100N_A \qquad (2)$$

where $M_{w,o}$ and M_w are the weight average molecular weights of the polymer initially and after receiving a total dose of D eV/g, the other symbols carrying the above mentioned meanings. Plots of $1/M_n$ and $1/M_w$ versus radiation dose D, yield the G values for crosslinking and scission. The determination of reliable values for G(S) and G(X) therefore depends on the precision and accuracy of the average molecular values used. One has to be careful with the operational and data analysis aspects of SEC since in addition to the inherent advantages of using a SEC chromatogram for the extraction of information on average molecular weights the same technique has some limitations.

The major limitation of SEC lies in the fact that it is a secondary method of molecular weight determination and requires calibration against a primary method to convert raw experimental data into molecular weight distributions. Studies of SEC separations have shown that polymers elute in reverse order of their hydrodynamic volumes. The early experimental studies of Benoit et al. (3) and the theoretical considerations Casassa (4) form the basis of of a universal calibration method, where [η] M Product is taken as a direct measure of the hydrodynamic volume of the polymer coil. When the [η] M product of a primary standard vs elution volume is constructed for a particular solvent, temperature and column system, the molecular weight averages and heterogeneity indexes of any polymer can be computed from this calibration curve by the Weiss and Cohn-Ginsberg method (5) using the following equations:

$$M_w = K^{-1/(a+1)} \sum_i w_i J_i^{\ 1/(a+1)} \qquad (3)$$

$$M_n = K^{-1/(a+1)} \left(\sum_i w_i J_i^{\ a/(a+1)} \right)^{1/a} \qquad (4)$$

$$HI = M_w/M_n \qquad (5)$$

In the above equations K and a are the Kuhn-Mark-Houwink-Sakurada (KMHS) constants of a polymer for a given SEC solvent and temperature. w_i represents the normalized peak heights of the chromatogram taken at equal small increments of elution volume and J_i is the universal calibration parameter which is defined as the product $[\eta]_i M_i$ where $[\eta]_i$ and M_i are the

intrinsic viscosity and the molecular weight of the ith species respectively.
The precision and accuracy of the molecular weight averages and distributions obtained by SEC are determined by the data analysis procedure and operational variables. The optimum working conditions for a SEC run are outlined in the detailed work of Mori (6). Even if the effects of improper working are minimised, uncertainties in the baseline and imprecision in the data will introduce serious errors into the SEC results. This has been investigated in detail by Rudin and co-workers (7) using a computer simulation.
To produce accurate molecular weight averages the chromatogram baseline must be stable and flat, the start and end of the peak must be clearly defined and the detector signal should not be noisy. Even if all these conditions are fulfilled there is still a serious source of error in the quantitative determinations by SEC. Since it is not practical or even possible to find primary standard samples for every polymer, the use of eqns.(3-5) given above is necessary for quantitative determinations by SEC. The precision and accuracy of the K and a values used in these equations will finally determine the reliability of SEC results. In the following paragraphs it is intended to show how serious is the error introduced in the molecular weight averages even when very small amount of uncertainties involved in the K and a values.
The KMHS constants for a given polymer in a particular solvent-temperature system vary considerably. Even for very narrow distribution polystyrene samples the value of a ranges between 0.66 and 0.77 (8) which means that an uncertainty of ± 10% exists around the mid-value. Similar differences are also observed in the K values and the subsequent calculations are therefore based on a ± 10% deviation in both a and K values. By using the KMHS constants of a = 0.768 and K = 6.09 $\times 10^{-5}$ of polystyrene in tetrahydrofuran (9) the primary calibration curve constructed for PS was converted into a universel calibration curve by following the method of Benoit and co-workers. The chromatograms of various poly(ethyl methacrylate), (PEMA), samples given in Table 1 were evaluated and M_w, M_n, and HI values were computed using eqns.(3-5). The KMHS constants of PEMA in THF used in these calculations were taken from the work of Samay et al. (9) as a = 0.679 and K = 15.49 $\times 10^{-5}$. Assuming these constants as the most reliable, the molecular weight averages and heterogeneity indexes calculated therefrom are similarly taken to be the most accurate values obtainable. After the determination of these values the same chromatograms were
Table 1 Characteristics of polyethyl methacrylate (PEMA) samples used in

this work

Polymer sample	$M_w x10^{-5}$	$M_n x10^{-5}$	M_w/M_n
PEMA F1	34.5	9.7	3.55
LPEMA F1	11.0	5.1	2.13
PEMA F2	7.6	3.9	1.95
LPEMA FO	5.3	1.9	2.80

reevaluated by changing the a and K values systematically around the original values by giving positive and negative deviations to KMHS values used in eqns. (3-5).

Following this procedure the effect of uncertainities in K values on M_w, M_n and HI values was determined for the PEMA samples but only the results pertaining to PEMA F1 is given in Table 2 since the results obtained for the remaining samples did not differ greatly from those listed. The relative errors introduced in weight and number average molecular weights resulting from deviations in K values are not significant.

Table 2 Errors introduced in Mw, Mn and HI values resulting from deviations in K values

% Deviations in K values	-12.8	-9.6	-6.4	-3.2	3.3	6.5	9.8	13.0
Corresponding % errors in								
Mw	8.5	6.2	4.0	1.9	-1.9	-3.7	-5.4	-7.0
Mn	8.5	6.2	4.0	1.9	-1.9	-3.7	-5.4	-7.0
HI	0	0	0	0	0	0	0	0

The uncertainities in the KMHS exponent a, however, cause significant errors in the calculations leading to M_w, M_n and HI. The errors in the molecular weight depend on the type of average taken and on the breadth of initial molecular weight distribution. The errors in the M_w, M_n and HI values of PEMA samples are listed in Tables 3-5. As is evident from Tables 3 and 4 even deviations of ± 3 % in a values lead to an almost ± 20 % error in M_w and ± 15 % error in M_n values. Therefore since deviations of ± 3 % in the KMHS exponent are very probable, the errors to be introduced due to uncertainties in a cannot be neglected. From the results given in Tables 3 and 4 it can be seen that the higher the molecular weight of the sample, the higher the error in M_w and M_n values for a given deviation in the KMHS exponent.

Table 3 % Errors in Mw values

	% Deviations in a values					
Sample	-8.7	-5.7	-2.8	3.1	6.0	9.1
PEMA F1	78.2	45.8	19.9	-17.8	-31.4	-42.6
LPEMA F1	69.2	40.9	17.9	-16.3	-29.1	-39.6
PEMA F2	66.5	39.5	17.4	-15.8	-28.3	-38.7
LPEMA O	66.3	39.3	17.3	-15.8	-28.2	-38.6

Table 4 % Errors in Mn values

	% Deviations in a values					
Sample	-8.7	-5.7	-2.8	3.1	6.0	9.1
PEMA F1	62.6	37.4	16.5	-15.2	-28.5	-38.9
LPEMA F1	59.8	35.8	15.9	-14.7	-26.4	-36.3
PEMA F2	58.4	35.2	15.6	-14.5	-26.0	-35.8
LPEMA 0	53.5	32.3	14.4	-14.5	-24.4	-33.9

Table 5 % Errors in HI values

	% Deviations in a values					
Sample	-8.7	-3.7	-2.8	3.1	6.0	9.1
PEMA F1	9.0	6.1	2.9	-2.9	-5.7	-8.2
LPEMA F1	5.9	3.7	1.8	-1.8	-3.5	-5.2
PEMA F2	5.1	3.3	1.5	-1.8	-3.1	-4.5
LPEMA 0	8.3	5.3	2.5	-2.6	-4.8	-7.0

It is clear that KMHS constants play a very important role in quantitative determinations by SEC. One must be very careful in selecting and using the

KMHS constants in quantitative analysis by SEC whenever universal calibration method is used. One possible solution to this drawback can be the use of broad molecular weight calibration method but even in that case the disadvantages inherent in that method brings about new problems. The best solution seems to couple a Low Angle Laser Light Scattering system with the Size Exclusion Chromatograph and obtain directly molecular weight distribution not size distribution.

CROSSLINKING

Since the major requirement for the analysis of a polymeric material by SEC is its solubility in the eluting solvent, highly crosslinked insoluble polymers can not be investigated by this method. Although Malihi et al. (10) were able to determine the gel content of acrylic latexes by SEC, the number of applications of this technique for direct determination of fully crosslinked structures is very low. As it is briefly mentioned in the previous section the change in the molecular size distribution of a lightly crosslinked polymer can be very easily followed by this technique. In a series of papers published by Keller and Patel (1), they were able to show that crosslinking takes place only in the amorphous regions when polyethylene is subjected to high energy radiations.

It is known that the molecular chains in semicrystalline polymers are either in fully extended forms or they fold back and forth in integral number of times to form the crystalline regions. These crystalline regions which are also called lamallae are bounded by amorphous regions. The amorphous regions can be eliminated by etching either with concentrated nitric acid or ozone. SEC chromatograms obtained for polyethylene samples irradiated before and after the elimination the amorphous regions yielded valuable quantitative information about the site of crosslinking, size of the crystal core, fold periodicity etc. In the following discussion we will show how SEC technique helps us to understand the effect of morphology, in addition to crystallinity, on the radiation induced crosslinking phenomena in poly (ethylene glycol).

Poly(ethylene glycol), PEG, is a model polymer for studying the influence of morphology on the physical properties of a polymer.PEG can be obtained in nearly monodispersed form and can be crystallized with the chains either

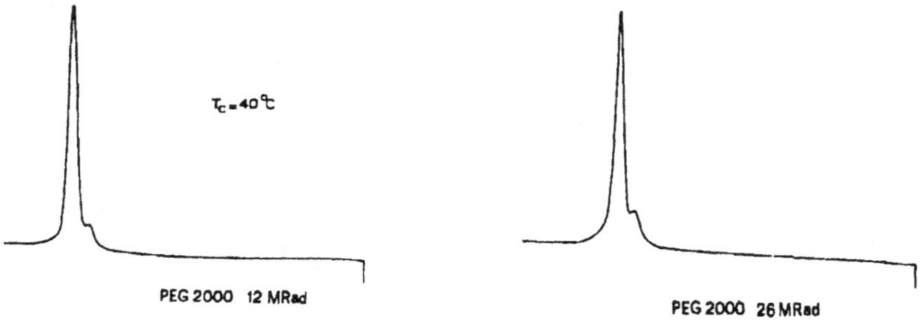

Fig. 1 SEC curves of hydroxy - terminated PEG 2000 irradiated to 12 and 26 Mrad doses.

fully extended or folded once, twice or three times etc. In the work described here PEG with molecular weight of 2000 crystallized at different temperatures in extended chain configurations were irradiated with gamma rays and SEC results are discussed(12). PEG samples crystallized at 40.0, 41.5, and 43.0 °C comprised 80, 68, and 60 % crystallinities. These samples were irradiated to various doses at room temperature in air and dissolved in SEC solvent immediately. Original sample had a heterogeneity index value of 1.04, therefore gave a single sharp peak. The chromatograms of PEG 2000 irradiated to 12 and 26 Mrad doses are given in Fig.1 where formation of distinct peaks at high molecular weight end of the curve is observed. The molecular weights corresponding to the shoulders appearing at lower elution volumes are almost exactly twice those associated with the sharp peaks. This doubling of original molecular weight implies that some of the PEG chains are linked at or near their end groups.

The doubling of molecular weight of PEG as observed here can be explained on the basis of end-to-end link formation. The orientation of hydroxyl end groups in the crystal-amorphous interface through hydrogen bonding facilitates end-to-end linkage. In order to convert the raw data given by the chromatograms in Fig. 1 into M_w and M_n values primary calibration curve was constructed (13). The reciprocals of these molecular weight averages are plotted against dose according to equations 1 and 2 and given in Figs. 2 and 3. The chain scission and crosslinking yields are computed simoultaneously from the slopes of these lines and the values obtained are

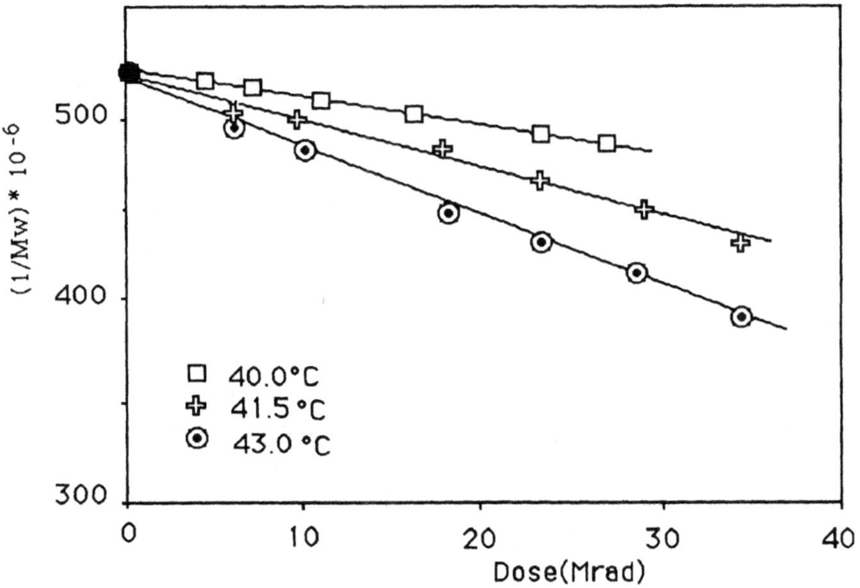

Fig. 2 Reciprocal Mw vs. Dose relations for gamma - irradiated PEG 2000 crystallized at different temperatures

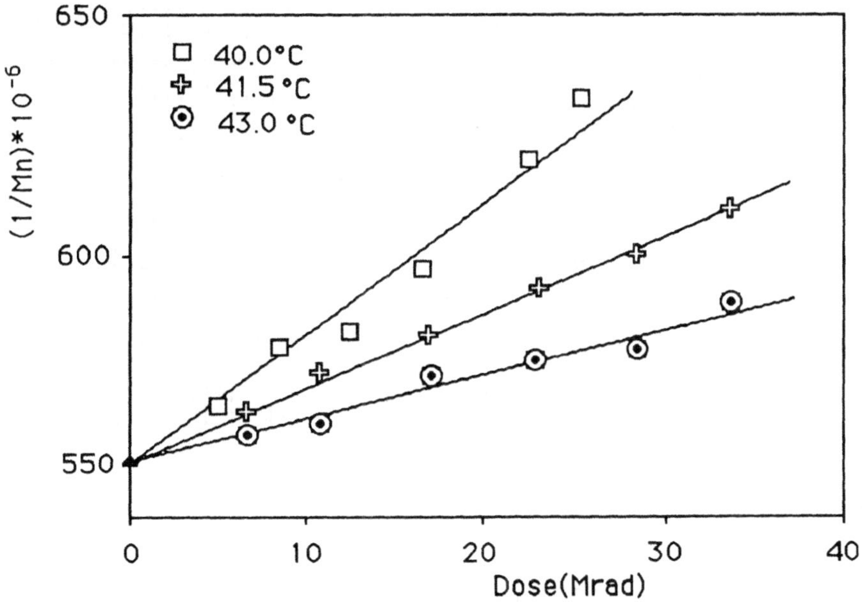

Fig. 3 Reciprocal Mn vs. Dose relations for gamma - irradiated PEG 2000 cyrstallized at different temperatures

listed in Table 6.

Table 6. Chain Scission and Crosslinking Yields for Hydroxyl-Terminated PEG

% Crystallinity	G (S)	G (X)	G(S)/G(X)
80	4.71	2.33	2.02
68	6.06	4.46	1.36
60	6.69	5.67	1.18

In addition to the above given quantitative results, SEC technique helps us to understand the mechanism of crosslinking process in the radiation crosslinking of PEG. The doubling of molecular weight of PEG was assumed to be as due to association taking place between the hydroxyl end groups of chains in the amorphous regions. The radicals produced at or migrated towards the chain ends favor the end-to-end linking because of the close proximity of the chain ends. If the formation of hydrogen bonding between

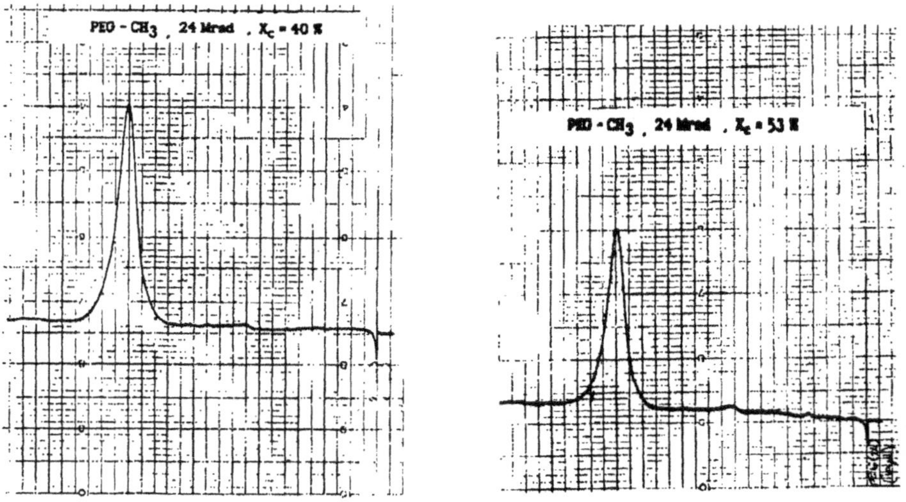

Fig. 4 SEC curves of methoxy - terminated PEG 2000 with differenet crystallinities, irradiated to 24 Mrad dose.

two PEG chains is one of the main factors responsible for enhanced crosslinking , then this assumption can be checked by replacing the chain ends of PEG with non-associogenic groups. For this purpose PEG with molecular weight of 2000 carrying methyl end groups was prepared, crystallized to definite % crystallinities and irradiated to various doses (14). The chromatograms of these samples taken at identical conditions with those of previous ones are given in Fig. 4. The major difference observed between the chromatograms shown in Figs. 1 and 4 is the disappearence of shoulder at the high molecular weight end of the chromatogram of hydroxy-terminated PEG. This shows that there is no longer any preference for the formation of end-to-end links and the crosslinking process becomes more random. This decrease in the extent of crosslinking is also substantiated by quantitative analysis of the chromatograms of methyl end-grouped PEG. The chain scission and crosslinking yields of two samples with different crystallinities are listed in Table 7.

Table 7. Chain Scission and Crosslinking Yields for Methyl-Terminated PEG

% Crystallinity	G (S)	G (X)	G(S)/G(X)
53	5.40	1.38	3.91
40	6.80	1.50	4.56

When the yields of chain scission and crosslinking for hydroxy- and methyl-ended PEGs are compared, it is seen that the presence of an associogenic group at the end of the polymer chains enhances the crosslinking. The effect produced, however, can be significant only with low molecular weight polymers. Nevertheless this study shows that when properly used and handled SEC technique can provide a wealth of information on the qualitative as well as quantitative aspects of chain scission and crosslinking processes.

REFERENCES

1) D. J. Harmon and H. L. Jacobs, J. Appl. Polym. Sci., 10, 253 (1976)

2) T. N.Bowmer, L. K. Cowen, J. H. O'Donnell, and D. J. Winzor, J. Appl. Polym. Sci., 24, 425 (1979)

3) H. Benoit, Z. Grubisic, P. Rempp, D. Decker and Z. G. Zillioux, J. Chim.

Phys., 63, 1507 (1966), and Z. Grubisic, P. Rempp and H. Benoit, J. Polym. Sci., Part B, 5, 753 (1967)

4) E. F. Casassa, J. Polym. Sci., Part B, 773 (1967)

5) A. R. Weiss and E. Cohn-Ginsberg, J. Polym. Sci., Part B, 7, 379 (1969)

6) S. Mori, J. Appl. Polym. Sci., 21, 1921 (1977)

7) W. J. Tchir, A. Rudin and C. A. Fyfe, J. Polym. Sci. Polym. Chem. Ed. , 20, 1443 (1982)

8) V. Deniz and O. Güven, Brit. Polym. J., 18, 112 (1986)

9) G. Samay, M. Kubin and J. Podesva, Angew. Makromol. Chem., 72, 185 (1978)

10) F. Malihi, C.-Y. Kuo, and T. Provder, J. Liq. Chrom., 6, 667 (1983)

11) G. N. Patel and A. Keller, J. Polym. Sci. , Polym. Phys. Ed. , 13, 303 (1975) , ibid, 13, 323 (1975)

12) O. Güven, E. Arca and J. Silverman , Proc. 6th Tihany Symp. on Radiation Chemistry, Balatonszeplak, Hungary, 517 (1986)

13) O. Güven, Brit. Polym. J. , 18, 391 (1986)

14) O. Güven, F. Yiğit and M. Şen, 7th Int. Meet. on Radiation Processing, Noordwijkerhout, The Nederlands, (1989)

INDEX

GPSR Compliance

The European Union's (EU) General Product Safety Regulation (GPSR) is a set of rules that requires consumer products to be safe and our obligations to ensure this.

If you have any concerns about our products, you can contact us on ProductSafety@springernature.com

In case Publisher is established outside the EU, the EU authorized representative is:

Springer Nature Customer Service Center GmbH
Europaplatz 3
69115 Heidelberg, Germany

Batch number: 10761599

Printed by Printforce, the Netherlands